CAMBRIDGE LIBRARY COLLECTION
Books of enduring scholarly value

Darwin

Two hundred years after his birth and 150 years after the publication of 'On the Origin of Species', Charles Darwin and his theories are still the focus of worldwide attention. This series offers not only works by Darwin, but also the writings of his mentors in Cambridge and elsewhere, and a survey of the impassioned scientific, philosophical and theological debates sparked by his 'dangerous idea'.

A Preliminary Discourse on the Study of Natural History

William Swainson FRS, was recognised principally as a zoologist, an ornithologist and a skilled and prolific illustrator. He also had a tremendous enthusiasm for seeking and identifying new species. In this 1834 volume however, Swainson addressed the nature of, foundations for and successful pursuit of zoology. It argues firmly for the key importance of taxonomy. Swainson was an ardent advocate of MacLeay's now entirely outmoded 'quinary' system of classification – even then a distinctly minority view. This sought affinities, patterns and analogies among organisms, in order to discern God's order. More than a mere curiosity, such work was of pivotal concern to enterprising naturalists of the 1820s and 1830s – including the young Charles Darwin. It also reached Robert Chambers, whose 1844 Vestiges of the Natural History of Creation was an important landmark in the development of the theory of evolution. 'We are indebted for what we know of these beautiful analogies to three naturalists – Macleay, Vigors, and Swainson, whose labours tempt us to dismiss in a great measure the artificial classifications hitherto used, and make an entirely new conspectus of the animal kingdom, not to speak of the corresponding reform which will be required in our systems of botany also.' – Robert Chambers, Vestiges of the Natural History of Creation (1844), p. 238

Cambridge University Press has long been a pioneer in the reissuing of out-of-print titles from its own backlist, producing digital reprints of books that are still sought after by scholars and students but could not be reprinted economically using traditional technology. The Cambridge Library Collection extends this activity to a wider range of books which are still of importance to researchers and professionals, either for the source material they contain, or as landmarks in the history of their academic discipline.

Drawing from the world-renowned collections in the Cambridge University Library, and guided by the advice of experts in each subject area, Cambridge University Press is using state-of-the-art scanning machines in its own Printing House to capture the content of each book selected for inclusion. The files are processed to give a consistently clear, crisp image, and the books finished to the high quality standard for which the Press is recognised around the world. The latest print-on-demand technology ensures that the books will remain available indefinitely, and that orders for single or multiple copies can quickly be supplied.

The Cambridge Library Collection will bring back to life books of enduring scholarly value (including out-of-copyright works originally issued by other publishers) across a wide range of disciplines in the humanities and social sciences and in science and technology.

A Preliminary Discourse on the Study of Natural History

William Swainson

CAMBRIDGE UNIVERSITY PRESS

Cambridge, New York, Melbourne, Madrid, Cape Town, Singapore,
São Paolo, Delhi, Dubai, Tokyo

Published in the United States of America by Cambridge University Press, New York

www.cambridge.org
Information on this title: www.cambridge.org/9781108005234

© in this compilation Cambridge University Press 2009

This edition first published 1834
This digitally printed version 2009

ISBN 978-1-108-00523-4 Paperback

This book reproduces the text of the original edition. The content and language reflect
the beliefs, practices and terminology of their time, and have not been updated.

Cambridge University Press wishes to make clear that the book, unless originally published
by Cambridge, is not being republished by, in association or collaboration with, or
with the endorsement or approval of, the original publisher or its successors in title.

THE

CABINET CYCLOPÆDIA.

CONDUCTED BY THE

REV. DIONYSIUS LARDNER, LL.D. F.R.S. L. & E.
M.R.I.A. F.R.A.S. F.L.S. F.Z.S. Hon. F.C.P.S. &c. &c.

ASSISTED BY

EMINENT LITERARY AND SCIENTIFIC MEN.

Natural History.

A PRELIMINARY DISCOURSE

ON

THE STUDY OF NATURAL HISTORY.

BY

WILLIAM SWAINSON, ESQ. A.C.G.
HONORARY MEMBER OF THE CAMBRIDGE PHILOSOPHICAL SOCIETY,
AND OF SEVERAL FOREIGN ACADEMIES.

LONDON:
PRINTED FOR
LONGMAN, REES, ORME, BROWN, GREEN, & LONGMAN,
PATERNOSTER-ROW;
AND JOHN TAYLOR,
UPPER GOWER STREET.

1834.

"THE WORLD CANNOT SHOW US A MORE EXALTED CHARACTER THAN THAT OF A TRULY RELIGIOUS PHILOSOPHER, WHO DELIGHTS TO TURN ALL THINGS TO THE GLORY OF GOD; WHO, IN THE OBJECTS OF HIS SIGHT, DERIVES IMPROVEMENT TO HIS MIND; AND, IN THE GLASS OF THINGS TEMPORAL, SEES THE IMAGE OF THINGS SPIRITUAL."

JONES OF NAYLAND.
The Fairchild Discourse for 1784.

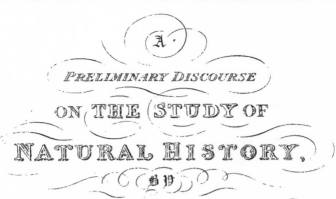

A PRELIMINARY DISCOURSE ON THE STUDY OF NATURAL HISTORY,

BY WILLIAM SWAINSON ESQ^R.

London:

THE CABINET
OF
NATURAL HISTORY.

CONDUCTED BY THE

REV. DIONYSIUS LARDNER, LL.D. F.R.S. L. & E.
M.R.I.A. F.R.A.S. F.L.S. F.Z.S. Hon. F.C.P.S. &c. &c.

ASSISTED BY

EMINENT SCIENTIFIC MEN.

A PRELIMINARY DISCOURSE

ON

THE STUDY OF NATURAL HISTORY.

BY

WILLIAM SWAINSON, ESQ. A.C.G.
HONORARY MEMBER OF THE CAMBRIDGE PHILOSOPHICAL SOCIETY,
AND OF SEVERAL FOREIGN ACADEMIES.

LONDON:

PRINTED FOR
LONGMAN, REES, ORME, BROWN, GREEN, & LONGMAN,
PATERNOSTER-ROW;

AND JOHN TAYLOR,
UPPER GOWER STREET.

1834.

CONTENTS.

PART I.

RISE AND PROGRESS OF ZOOLOGY.

Preliminary Observations. — Division of the Subject. — First Epoch. — Aristotle. — Pliny. — Second Epoch. — Rondeletius.— Gesner. — Aldrovandus. — Mouffet.— Topsal. — Maurice of Nassau. — Marcgrave. — Piso. — Merrett. — Goedartius. — Redi. — Swammerdam. — Lister. — General Remarks on the Era of Willughby and Ray. — Grew. — Pettiver. — Albin. — Sloane. — Seba. — Third Epoch. — Linnæus.—Ellis.—Linnæan School.—Rumphius. — D'Argenville.— Regenfuss.— Rœsel. — Edwards.— Trembley.— Gronovius. — Reaumur. — Comparison between Linnæus and Buffon.—Linnæan School.—Artedi.— Sulzer.— Sepp·—Scopoli. —Schœffer.—Hasselquist. — Osbeck.— Forskal. — Sparrman.— Pennant.—White.— Drury.— Martini and Chemnitz.—Wilks.— Fabricius. — Thunberg.— Müller.— Forster.—Villers.—Schrank.— Moses Harris.— Cramer.— Stoll.—Schreber.—Pallas.— Schroeter.— Born.— Merrem. —Hermann.— Bloch.— Schneider.— Schœpf.—Latham.— Shaw. — Sir J. Smith. — Berkenhout. — Lewin. — Otho Fabricius. — Olivi. — Entomological Illustrative Works of this Period. — Ernst. — Esper. — Hübner. — Herbst. — Jablonsky.—Voet.—Wolf. — Minor Writers — Panzer. — Petagni. — Rossi. — Paykull. — Lespeyres. — Gmelin. — Buffon's School. — Planches Enluminées. — Bonnet. — De Geer.— Brisson.— Adanson.—Duhamel.—Sonnerat.— Sonnini. — Levaillant. — Fuessly. — The Modern French School. — Cuvier. — Discovery of the Circular Nature of Affinities - - Page 1

PART II.

ON THE GENERAL NATURE AND ADVANTAGES OF THE STUDY OF NATURAL HISTORY.

CHAP. I.

Introductory Remarks. — What Natural History is, — in a general Sense, and as now restricted. — Division of the Subject. — Reflections on Nature and Art. — Distinctions, and Object of the Study - - - Page 93

CHAP. II.

Natural History viewed in its Connection with Religion. — As a Recreation. — As affecting Commerce and the economic Purposes of Life. — As important to Travellers - 107

PART III.

OF THE PRINCIPLES ON WHICH NATURAL HISTORY RELIES FOR ITS SUCCESSFUL PROSECUTION, AND THE CONSIDERATIONS BY WHICH THE NATURAL SYSTEM MAY BE DEVELOPED.

CHAP. I.

On the Dismissal of Prejudice - - - 152

CHAP. II.

On the Principles on which Natural History, as a Branch of Physical Science, is to be studied - - - 165

CHAP. III.

On Arrangements generally; and on those Considerations which should form the Basis of every Attempt to classify Objects according to the System of Nature - - 188

CHAP. IV.

On Theories in general; and on the Modes and Considerations by which they are to be verified - Page 201

CHAP. V.

On the Characters of Natural Groups - - 236

CHAP. VI.

On the Importance of Analogy when applied to the Confirmation of Theory - - - - 282

PART IV.

ON THE PRESENT STATE OF ZOOLOGICAL SCIENCE IN BRITAIN, AND ON THE MEANS BEST CALCULATED FOR ITS ENCOURAGEMENT AND EXTENSION.

CHAP. I.

Introductory Remarks. — Some Account of the Nature and present State of our Scientific Societies and Institutions, and of the Means they possess of encouraging Science. — National Encouragement - - 296

CHAP. II.

On the National Patronage of Science in other Countries, as compared to its Neglect by the British Government. — The Causes which produce this Neglect, and the Expediency of removing them - - - 339

CHAP. III.

On the Means possessed by the Government and Universities for protecting and encouraging Science. — On Titular Honours - - Page 367

CHAP. IV.

Suggestions for the Reform and Improvement of our Scientific Societies - - - - 428

PRELIMINARY DISCOURSE

ON

THE STUDY

OF

NATURAL HISTORY.

PART I.

ON THE RISE AND PROGRESS OF ZOOLOGY.

PRELIMINARY OBSERVATIONS. — DIVISION OF THE SUBJECT. — FIRST EPOCH. — ARISTOTLE. — PLINY. — SECOND EPOCH. — RONDELETIUS. — GESNER. — ALDROVANDUS. — MOUFFET. — TOPSAL. — MAURICE OF NASSAU. — MARCGRAVE. — PISO. — MERRETT. — GOEDARTIUS. — REDI. — SWAMMERDAM. — LISTER. — GENERAL REMARKS ON THE ERA OF WILLUGHBY AND RAY. — GREW. — PETTIVER. — ALBIN. — SLOANE. — SEBA. — THIRD EPOCH. — LINNÆUS. — ELLIS. — LINNÆAN SCHOOL. — RUMPHIUS. — D'ARGENVILLE. — REGENFUSS. — RŒSEL. — EDWARDS. — TREMBLEY. — GRONOVIUS. — REAUMUR. — COMPARISON BETWEEN LINNÆUS AND BUFFON. — LINNÆAN SCHOOL. — ARTEDI. — SULZER. — SEPP. — SCOPOLI. — SCHŒFFER. — HASSELQUEST. — OSBECK. — FORSKALL. — SPARRMAN. — PENNANT. — WHITE. — DRURY. — MARTINI AND CHEMNITZ. — WILKS. — FABRICIUS. — THUNBERG. — MÜLLER. — FORSTER. — VILLERS. — SCHRANK. — MOSES HARRIS. — CRAMER. — STOLL. — SCHREBER. —

PALLAS. — SCHROETER. — BORN. — MERREM. — HERMANN. — BLOCH. — SCHNEIDER. — SCHŒPF. — LATHAM. — SHAW. — SIR J. SMITH. — BERKENHOUT. — LEWIN. — OTHO FABRICIUS. — OLIVI. — ENTOMOLOGICAL ILLUSTRATIVE WORKS OF THIS PERIOD. — ERNST. — ESPER. — HÜBNER. — HERBST. — JABLONSKY. — VOET. — WOLF. — MINOR WRITERS. — PANZER. — PETAGNI. — ROSSI. — PAYKULL. — LESPEYRES. — GMELIN. — BUFFON'S SCHOOL. — PLANCHES ENLUMINÉES. — BONNET. — DE GEER. — BRISSON. — ADANSON. — DUHAMEL. — SONNERAT. — SONNINI. — LEVAILLANT. — FUESSLY. — THE MODERN FRENCH SCHOOL. — CUVIER. — DISCOVERY OF THE CIRCULAR NATURE OF AFFINITIES.

(1.) To form a just estimate of the relative position of any science at a given period, it is necessary that the prominent events in its history be rightly understood. It seems, therefore, expedient to commence this discourse with a slight sketch of the rise and progress of zoological science; or, more properly, of the progressive discovery of the forms, structures, and habits belonging to the animal world; a world replete with such an infinity of beings, each possessing so many peculiarities of habit and economy, that, notwithstanding the united efforts of human research for thousands of years, there is not one of them whose history, as yet, can be pronounced complete.

(2.) The vast and diversified field of enquiry over which zoology extends, and the many distinct portions into which it is now distributed, render it extremely difficult to embrace the whole in one general exposition. For it has happened, that at one period of time while our knowledge has made gigantic progress in one department, it has been stationary, or even retrograde, in others; and at another epoch we find that original research has been abandoned, and the technicalities of system and nomenclature alone regarded. To meet the first difficulty, and to preserve, nevertheless, a connected narrative, it seems advisable to treat the subject historically; and pre-supposing certain epochs in this science, to detail the peculiar characteristics of each. This will of course lead to some enquiry into the merits of those who have successively promoted or retarded the progress of knowledge; or who have been the founders of systems and methods, which for a time have endured, and then been laid aside. The revolutions of science are almost as frequent, and often more extraordinary, than those of political institutions. Both are results, not so much of the talents or efforts of large communities acting simultaneously, as of the influence of some one individual, whose qualities, good or bad, have not unfrequently worked the overthrow of laws, and modes of thinking, which had long been supported by the voice of a nation. It is, therefore, the part of the natural not less than of the political historian, to trace the causes of such revolutions, as far as possible, to their sources; and not to rest contented with the bare enumeration of the facts themselves, or of the results which followed.

(3.) Nor is the above the only difficulty of the task before us. To estimate aright the progress of this science, it is essential to draw a just distinction between analogical research and systematic arrangement; or, in other words, between the minute investigation of the properties and characters of an animal, and its subsequent arrangement among other animals. It has been the misfortune of those who have written — in some respects ably — upon the rise and progress of zoology, that this distinction has either not been perceived, or has been entirely set aside. Hence it has resulted that praise and blame have been frequently misapplied; while discoveries of the highest interest have been quite overlooked in the fancied importance attached to the maker of a system, or to the industry of a nomenclator. Without, at present, entering further into these essential differences between the labours of naturalists, we must bear in mind that all true knowledge of the laws of natural combination takes its rise from minute analysis; and that the value of a system is to be judged of according to the degree with which it arranges in harmonious order, all the various and infinitely diversified facts resulting from analysis. Of artificial systems there may be no end, because the materials of which they are composed show a diversity of relations: each system may differ from the other, yet each may have something to recommend it. But with the materials employed for their construction the case is quite different: the analysis of a species, if correctly made, remains for ever, unchangeable and unchanged: it is permanent; it cannot be gainsaid, nor does it perish with the system into which it may be incorporated. The system may be overthrown, yet the analysis remains. True it is that minute research is of more easy accomplishment than the power of generalising: the one requires only a simple accuracy of observation, the other an enlarged and comprehensive judgment. But, when once a system, like that either of Aristotle or of Linnæus, has been framed, it is easy for a host of imitators to follow, each making some fresh modifications, or some small improvements upon the models before him; and thus dazzling the world with *a new system*, which the inventor would never have composed, had he been left to his own unassisted powers of combination. In estimating, therefore, the respective merits of the two classes of naturalists here alluded to, we shall be obliged to assign a much lower station to some names than has been done by our predecessors, and transfer that praise which has been bestowed upon them to others whose labours, although less brilliant, have more contributed to the advancement of science.

(4.) In reference to the above observations, we shall now take a rapid sketch of the history of zoology under the following epochs: — 1. Its foundation by Aristotle; 2. From the revival of learning to the time of Linnæus; and, 3. From the appearance of the Systema Naturæ of Linnæus, to that of the Règne Animal by Cuvier.

(5.) The state of natural history, in the early ages of the world, must ever remain more a matter of conjecture and of theory than of positive fact. Some acquaintance with the properties of animals was certainly possessed by our first parents, who were enabled, by the Divine agency, to assign names to the beasts of the field, and to distinguish such as were adapted to their wants. The wisdom of the wisest of men, also, was extended to the works of that God whom he worshipped; but these and similar ntimations in ancient history, whether sacred or profane, must not be interpreted too literally, or be supposed to imply more than that the knowledge of natural history, possessed by the early inhabitants of the earth, was commensurate with what was known of astronomy or other of the physical sciences.

(6.) Passing over, therefore, those obscure ages, when all human learning was in its infancy, we may date the rise of zoology, as a study, from the time when the immortal Aristotle directed the powers of his mind to the animal world ; and in his famous book, Περι Ζωῶν ‘Ιστοριας, first sought to define, by the precision of language, those more prominent and comprehensive groups of the animal kingdom, which, being founded on nature, are exempt from the influence of time and the mutability of learning. Had this extraordinary man left us no other memorial of his talents than his researches in zoology, he would still be looked upon as one of the greatest philosophers of ancient Greece, even in its highest and brightest age. But when it is considered that his eloquence, and his depth of thought, gave laws to orators and poets, — that he was almost equally great in moral as in physical science, and that no department of human learning escaped his research, or was left unilluminated by his genius, — we might be almost tempted to think that the powers of the human mind, in these latter days, had retrograded; and that originality of thought, and of philosophic combination, existed in a far higher degree among the heathen philosophers than in those who followed them. A moment's reflection, however, will show that such ideas are grounded upon partial considerations, and they are at once refuted by such names as those of Newton and Bacon. Furthermore, it should be remembered that the most ordinary observer can readily distinguish a quadruped from a bird, a snake from a fish, and a vertebrated from a boneless animal. All these distinctions are obvious, and, therefore, known even to the vulgar. Nor does it require any great skill to express these differences in words. The same may be said of those secondary divisions by which a beetle may be known from a butterfly, and these, again, from a bee. It is not so much, therefore, from having embodied facts like these into classic language that the philosopher of Stagyra derives his high fame; it rather reposes upon the peculiar tact with which he brought the rules of philosophic reasoning to bear upon a subject hitherto neglected, — upon the extent and depth of his personal researches, — upon the clearness with which he arranged his results, — and, above all, upon those obscure perceptions which he acquired, while so employed, of hidden truths, which were only to be developed in subsequent ages. Nor should that innate grandeur of his mind be forgotten, which led him, in an age of universal superstition, to discard from his work all those popular tales, and fancies, and beliefs, which were received by the mass of his countrymen as religious truths, sanctioned by antiquity, interwoven in their history, and consecrated in their poetry. The death of this great father of our science was the death of natural history in the Grecian era. The splendour of his discoveries passed like a comet. He left no luminary behind to follow in his wake, still less to throw additional light upon realms which he had but glanced upon. From the decline of Grecian learning until its partial revival among the semi-barbaric Romans, a long interval of darkness intervened; and it was only after a lapse of nearly 400 years that we find a solitary philosopher—the elder Pliny—calling the attention of his countrymen to the wonders of nature, and following up the pursuits of the Grecian sage. The Roman naturalist strove to follow in the path of his great predecessor; for, like him, he undertook to illuminate the whole empire of science and of learning: but he had neither the erudition nor the genius requisite for his gigantic project. His voluminous works rather show us a compilation of other men's thoughts and discoveries, than a selection of well digested information, or of original research. We find the wheat intermixed with an abundance of chaff: the nutritive grain and the useless straw are equally hoarded, and brought into the garner. Amidst all the polished graces of diction, great and diversified erudition, and no inaptitude for occasionally describing with clearness and precision, we look in vain for the powerful genius and the originality of thought of his great master, and we at once perceive that natural history, or rather zoology, under the Romans, had made a retrograde movement. The powerful mind of Aristotle, which led him to reject with disdain the credulous tales and fabulous stories of the age, can nowhere be traced in the writings of Pliny, whose works, on the contrary, abound in fables and in prodigies, at once manifesting that weakness of mind inseparable from credulity, or that disinclination to investigate truth, which is the sure mark of a secondary order of intellect. It is difficult to account for this paucity of original information and abundance of fable in the writings of Pliny, seeing that he lived in an age when Rome might be said to have possessed the most magnificent menagerie the world ever witnessed. Her barbarous exhibitions of animal combats, — conducted on a scale of savage splendour, which almost shakes our credulity,—assembled within her walls fresh supplies of hundreds of living animals, collected from all the regions over which her empire extended, and augmented by the forced or voluntary contributions of those allies who sought the protection or friendship of the mistress of the world. These menageries were not only filled with lions and other ferocious animals, destined for the circus, but contained, in all probability, whatsoever was rare or curious among the more peaceable tribes; since these creatures frequently formed a conspicuous feature in triumphal processions, and were no doubt taken care of afterwards.* The Camelopardalis of northern Africa (C. *antiquorum*, Sw.) was well known to the Romans; but that of the southern regions, we may presume, was too far removed from their empire. Certain it is, however, that of all these advantages

---

* Pliny himself is the authority for these facts. He informs us that Quintus Curtius first began the custom. Scylla exhibited the terrific spectacle of a combat of 100 male lions; but this savage amusement was far outdone by Pompey, who assembled at one time no less than 600 of these beasts. Cæsar, also, had one of 400.

Pliny made but little effective use. If any further proofs were requisite to show the declension of natural history under the Romans, it would only be necessary to cite the fables and absurdities of Ælian, and one or two others, with whom expired all records of the science for nearly 1400 years.

(7.) The second era of our history commences with the revival of learning in the sixteenth century, and terminates with the institution of system by our celebrated countrymen Lister, Willughby, and Ray. It is difficult to trace the first dawn of natural history during this period, or to ascertain which was the first printed book that treated on the nature of animals. The *Ortus Sanitatis**, printed in 1485, a most curious and exceedingly rare book, is the earliest we have seen; and, to judge from the grotesque rudeness of its figures, was, perhaps, one of the very first attempts to represent animals by wood-cuts. Passing over, however, this and similar memorials of a dark age, the first writer who really deserves notice is Belon of Mans, who was born in 1517, and who seems to have made the history of birds his exclusive study. He may not have been the first writer on natural history, in regard to priority, since the revival of science, but he was most assuredly the first who treated the subject with any regard to system; and when we consider the unenlightened era in which he lived, and the diffi-

---

* Ortus Sanitatis. De herbis et plantis, de animalibus et reptilibus, de avibus et volatilibus, de piscibus et natatilibus, de lapidibus, &c. 1485. Small folio. Ascribed by some to a Doctor Cuba.

culties he must have had to contend with, he may justly be considered the reviver of natural history in the sixteenth century. Belon, it is true, seems to have paid very little attention to Aristotle, and to have been totally ignorant of the philosophy of his subject; yet his arrangement, so far from being despicable, is much more natural than has been generally supposed: this will at once appear from a glance at his system. Commencing with the land birds of prey, as the vultures, falcons, shrikes, and owls, he passes to the water birds of prey, as the cormorant, albatross, &c.; the wading order naturally follows, and from this he proceeds to the gallinaceous tribe, including the ostrich family. The two last chapters are devoted to the pigeons, crows, and thrushes, and all the smaller perching birds. Now if we look to this arrangement, not in regard to its details, but to the general character of its primary groups, we have, in fact, precisely the same disposition as that which we now know to be the natural series. Here we find the modern orders of *Raptores, Natatores, Grallatores, Rasores,* and *Insessores* following each other in the order of their true affinities, and exhibiting the circular disposition of the whole feathered creation. The chief objection to Belon's arrangement is to be found in his details, where he places not only the plovers but the larks and buntings within his gallinaceous division, instead of associating them with his perching families. But what more could be expected in the infancy of science, and from the first who gave to it a definite form? In this branch of zoology, therefore, Belon must be considered as much the master of Willughby, as the latter was of Linnæus. He was, moreover, a perfectly original writer, and described his subjects, considering the age in which he flourished, with remarkable exactness. As he was the first systematist of this period, so he was the only eminent writer on birds between Aristotle and Ray; while the manner in which he treated his subject showed a mind much superior to his contemporaries in other branches.

(8.) Natural history seems to have again revived in the countries which gave it birth in ancient times, for nearly all the remaining writers of this period were natives of Italy, or at least of southern Europe. The year 1554 was remarkable for the appearance of two works on ichthyology, the one by Rondeletius*, (or, as the French write the name, Rondelet,) an early professor of medicine at Montpellier; the other by Salviani †, a physician of Rome. The first treats at great length on the nature of fish in general, and describes, with considerable exactness, a large number of those found in the Mediterranean. There is no attempt at systematic arrangement, yet the subjects are not promiscuously introduced; for the sharks, the eels, the rays, and other natural groups, are placed in distinct chapters. These being dismissed, our author proceeds to notice a variety of other animals belonging to different classes, merely, as it would appear, because they have something of the nature of fish by living in the sea. In his sixteenth chapter he accordingly jumbles

---

* Gulielmi Rondeletii. Libri de Piscibus Marinis, in quibus veræ Piscium effigies expressæ sunt. Lugduni, 1554.

† Hyppolyti Salviani de Citta di Castello. Aquatilium Animalium Historiæ Romæ, 1554.

together tortoises, whales, and seals; and concludes with giving us various descriptions and grotesque figures of certain marine monsters, designated as *de montro leonino, de pisce monachi habitu,* and *de pisce episcopi habitu.* Where he found the extraordinary originals from which these cuts were taken does not exactly appear: they were probably fabricated from the skins of some large species of shark or ray, by the ecclesiastics of that period, to attract the superstitious veneration of the populace, by persuading them that even the sea contained monks and bishops. The letter-press exhibits all the prolixity and cumbrous learning of the age, with abundant quotations from Aristotle, yet without the least spark of his philosophic spirit or of his arrangement. With all these defects, this early specimen of ichthyology has great and even extraordinary merit in the excellency of the wood-cuts copiously introduced in its pages: they are bold and accurate, and in general so characteristic, that nearly all the species may be at once identified. Salviani's work on the same subject appeared simultaneously with that of Rondeletius, both being printed, as before observed, in 1554; but the former is now very rare, and is not in our library: the figures, which are engraved upon copper, are generally mentioned as very good. While these two patriarchs of ichthyological science were directing their investigations to one branch of natural history, two other, equally zealous and more ambitious in their projects, were respectively labouring on a general history of all animals, influenced no doubt by the example of Pliny, whose work was more adapted to the mental comprehension and the credulity of the sixteenth century, and was doubtless held in much higher estimation than the masterly but too philosophic treatise of the great Aristotle. The laborious naturalists alluded to were Conrad Gesner* and Ulysses Aldrovandus; the first a physician of Zurich, the latter of a noble house of Bologna, in the university of which city he was a professor. Gesner was born in 1516, and died at the age of 49; so that it would seem he did not live to see the publication of his work, which was printed in three folio volumes at Frankfort, in the year 1585. He appears to have been an industrious compiler of other men's labours, adding little of his own, and quite destitute of all notions of system, the subjects being arranged alphabetically. This voluminous compendium is ornamented with wood-cuts of very unequal execution; some being very tolerable, others very bad. Aldrovandus, who must have been working at the same time (for he was born in 1525), dedicated his life and his fortune to a similar undertaking, still more diffuse and voluminous than the compilation of Gesner: he, likewise, lived not to see the publication of his work, which extended to no less than fourteen folio volumes, the greatest portion of which were printed after his death. The original descriptions in this work are more numerous and accurate than those of Gesner, but its author had little judgment and still less genius for his task: he collects from all quarters every thing that had been

---

* Conrade Gesneri Tigurini, Historia Animalium. Francofurti, 1585.

written upon animals, whether true or false; while the facts he must have had the power of verifying, and the fables that he might have detected and exploded, are confusedly mixed together, as if to swell the cumbrous folios upon which he had spent his life in compiling.* That both these works, however, were greatly instrumental in diffusing a taste for the study of nature is very apparent, even from the simple fact of their sale being so great as to induce the publishers of that period to incur the enormous expense of printing them. We question very much, whether any bookseller of the present age would undertake to bring out fourteen folio volumes upon natural history, even were they to contain the joint labours of all the eminent naturalists of the present age. While this great compilation, or rather encyclopædia, of zoological knowledge was in progress, Fabius Colonna, a physician of Rome, published two treatises on natural history, of a much higher character than those of his contemporaries, and which have procured for their author a high reputation from the moderns.

(9.) A taste for natural history had hitherto been confined to the Continent, but in the year 1634 it had at length reached England; and the *Theatrum Insectorum* of Mouffet came forth as the first zoological work ever printed in Britain. Mouffet appears to have been physician to the earl of Pembroke, and to have made insects his sole object of study: nor was he the only one who

---

* Ulysses Aldrovandus. Philosophi et Medici Bononiensis Historia Naturalium, in Gymnasio Bononiensi profitentis. Bononiæ, 1599—1640.

at this early period found pleasure in entomological recreations; for we find in the title-page the names of two congenial friends, Edward Wotton and Thomas Penn, associated with his own as fellow-labourers in the production of this curious volume. In a work of so early a date, we must not look for any great departure from the prolixity and credulity of contemporary writers, or any thing beyond the rude classification of separate groups into distinct caputs or chapters; overloaded, as was the fashion of the age, with heavy details of common truths, and obscured by a want of precision and by absurd fables, handed down by Pliny to all succeeding compilers. The contents, however, are so far digested as that the winged and the apterous orders form the two principal divisions of the work; but then, in the latter, the author treats of caterpillars and grubs as if they were insects arrived at maturity, and of genuine worms as if they also were insects: the wood-cuts are many, but of great inferiority, even for the period of their execution; and they show how tardy the progress in England had yet been of the fine arts. It is worth while observing, that, with the exception of Belon, all the authors we have yet spoken of published their works in Latin: this, in fact, was then the universal language of learning and of science. Knowledge was chiefly confined to the ecclesiastical and the medical professions, and to those few of the higher orders who had been educated by the clergy; but the mass of the people, even those of the gentry and middle classes, were profoundly ignorant. The religious establishments and the collegiate institutions were at the same time the national libraries; and few, not residing within their walls, ever dreamed of pursuing knowledge as a recreation. We may, therefore, fairly infer, that some taste for natural history had begun to show itself among the common people, when we find that in the year 1658 one Edward Topsel*, an ecclesiastic of St. Botolph's, London, published an English translation in folio, not only of Gesner's work upon quadrupeds, but also of the aforesaid Theatrum Insectorum of Dr. Mouffet; thus placing in the hands of our countrymen, in their own language, the two best works upon beasts and insects that had appeared since the revival of learning.

(10.) That a love of knowledge had found its way beyond the precincts of cloisters and the halls of professors was now evident; for it was about this time that our science was protected by one of the most remarkable men of his age,—the great and chivalrous Count Maurice of Nassau; a name which will be immortalised by the historian, no less than by the naturalist. In Count Maurice was united the accomplished statesman, the victorious general, and the munificent patron of science; and in each of these characters his merits were so high, that the history of modern times cannot afford us his parallel, unless it be found in the late Sir Stamford Raffles,— a name equally dear to our science, although perhaps not so brilliant for " battles won, and standards taken." Count Maurice, upon his assuming the

---

* Topsel's History of Four-footed Beasts and Serpents; to which is added, Mouffet's Theatre of Insects. London, 1658. 1 vol. folio.

command of the Dutch armament, which subsequently dispossessed the crown of Portugal of nearly all its Brazilian possessions, took with him, as if anticipating victory and subsequent ease, a young and enthusiastic naturalist, whom we now look on as the venerable Marcgrave, the father of Brazilian zoology. Not content with his aid, the count employed artists and botanists to draw, and collect, and preserve, every thing that might interest the naturalists of Europe. To this munificent patron was the learning of the seventeenth century indebted for the first account, ever published, of the natural history of tropical America. Considering the then state of science, Marcgrave's work, written probably when he was not more than twenty-five, abounds with a vast mass of new and original information, very different from what was to be found in the crude and verbose compilations of this period. Unfortunately, however, Marcgrave lived not to arrange and digest these materials, as he no doubt would have done had he returned to Europe. Anxious to extend his discoveries, he accompanied one of the bold expeditions of his patron to attack the Portuguese possessions on the coast of Guinea, where he fell a victim to the climate at the early age of thirty-four. His original MSS., and a collection of drawings chiefly of the rare fishes of Brazil, made by his accomplished patron, are said to be preserved in the Royal Library of Berlin. His work on Brazil was published, with those of Piso and Bontius on India, in 1768. Marcgrave had talents of a very high order; for, besides his zoological and botanical labours, he wrote on the customs of the natives, — studied and analysed their language, — made astronomical observations, and evinced, in short, the possession of all that varied knowledge which we should only look for in an accomplished traveller of this century. Such a man was worthy of so great a patron; and he must be considered as by far the most eminent naturalist of the era in which he lived. On looking back to the history of our science from the revival of letters, it will be observed that nearly all who had contributed to its advancement were little better than voluminous compilers; who, to a scanty stock of original information, superadded a ponderous load of ancient lore, gathered from the fables of Pliny, and the credulous writers who followed him: they seemed to think that the value of their works would be estimated by their bulk, or they were probably deterred from prosecuting original research, by their veneration for antiquity. This school, which had been founded by Pliny, seems to have expired with Mouffet; for in Marcgrave and Bontius we have the first specimens of local *faunæ*, or natural histories of particular regions. Our own country followed this example; for in 1667 appeared the Pinax of Dr. Merrett*, the first work that was devoted exclusively to the animals and plants of Great Britain. It is written in Latin; and although very curious in its way, is yet a very imperfect performance, for there are no specific names, and the

---

* Pinax Rerum Naturalium Britannicarum, continens Vegetabilia, Animalia, et Fossilia, in hac Insula reperta inchoatus. Lond. 1667.

descriptions are so short, that very few of the commonest insects can be identified.

(11.) While entomology was thus making a slow and painful progress in England, the science received a new impulse on the Continent from the *experiments*, as they were then termed, of two celebrated men, Goedartius and Redi: the one undertook to investigate the metamorphoses of insects, the other was chiefly occupied in tracing their vital functions, and both may be thus considered as the founders of zoological analysis. The little volumes of Goedart, printed in 1662, showed a very marked improvement in the entomology of the seventeenth century, not so much in the descriptions as in the faithfulness of the numerous copper-plates, representing the larva, pupa, and perfect insect of a considerable number of lepidoptera: these "experiments" are extended to many species of the other orders; and the lates are so good, that they may be consulted with advantage even in the present day.* It is curious to trace, even at this more advanced period, the remnant of that superstition regarding common animals which was so prevalent in the preceding century. Upon turning to such plates of the work before us as represent the angulated chrysali of butterflies, the reader will perceive them transformed into the likenesses of swarthed mummies, where the nose, eyes, and chin, are distinctly marked out: there is certainly a curious resemblance to the human

---

* Metamorphosis et Historia Naturalis Insectorum. Autore Joanne Goedartio, cum Comentariis D. Joannis de Mey. Medroburgi, 3 vols. The date only appears at the end of the dedication, " 27 Januarii, 1662."

head in these *coffins*, as they were called by our ancestors; and if this relation is looked upon not as a resemblance, but as an analogy, it is in perfect accordance with truth. The experiments of Goedartius obviously led the way to those of Redi, on the generation of insects*, published also in three small volumes in different years. On the value of their contents we know but little; for the work is not now before us. Nor is this to be regretted, for both these names were eclipsed by one who was then labouring in the same field of analysis; but gifted, in every respect, with far greater talents. This was the celebrated Swammerdam, who died, at the early age of forty-three, a worn-out martyr to laborious study. The limits of this sketch will not permit us to expatiate on the life and discoveries of this extraordinary man. Suffice it to say, that the lapse of nearly one hundred and fifty years has in no degree weakened the value of his anatomical discoveries; and that so far as his researches were prosecuted, he has not been excelled by the greatest comparative anatomists of modern times. All the great truths on the metamorphosis of insects originated from this laborious and indefatigable observer, who was unquestionably the master and the guide of Lyonnet, Roemer, Bonnet, and all those who subsequently pursued the same path. Swammerdam, in short, was the great father of analysis, as Aristotle was of philosophic generalisation; and although their excellencies are

---

* Francesco Redi. Experimenta circa Generationem Insectorum. Amstelodami, 1671, 1686, 1712.

of different degrees of merit, and have never perhaps been united in one person, yet they are both essential to the perfection of our science. Swammerdam sunk into an early grave, before he could give to the world the result of his labours; and but for the patriotic feeling and the munificent liberality of the great Boerhaave, his manuscripts, drawings, and engravings would probably never have seen the light. Were it not customary to date the different stages of this science from the periods when particular systems of arrangement were in vogue, we should consider that a new era was commenced by Swammerdam, rather than by Ray; for the one enriched science with a mass of important facts, entirely and absolutely new; while the other merely employed these facts to construct a system, and this system chiefly modelled from that already sketched out by Aristotle. The original edition of Swammerdam's incomparable volume, in Latin and Dutch, was published at Leyden, under the superintendence of Boerhaave, in 1738; and it at length obtained so much reputation, that an English translation* appeared twenty years afterwards. We have introduced the name of Swammerdam in this part of our history because he was the contemporary of Goedart, Merrett, and Lister; and was prosecuting his researches at the same time, although science received no benefit from his discoveries until many years after.

(12.) Resuming, therefore, the thread of our narrative from the rude performance of Merrett,

---

* The Book of Nature; or, the History of Insects. By John Swammerdam. Translated by Thomas Floyd. Revised, &c. by John Hill, M.D. London, 1758. 1 vol. folio.

in his Pinax, published in 1667, we find that the next publication of any moment bears the name of one of the great naturalists of this era, — Dr. Martin Lister, secretary to the Royal Society (then but recently instituted), and chief physician to queen Anne. The first work of this father of conchology makes known the spiders, the shells, and the fossil echini, &c. of Great Britain; all of which are not only well described, but are accompanied by tabular systematic arrangements, superior to any that had yet been framed, and fully equal to those subsequently given by Ray. Lister, in fact, is unquestionably the inventor of system; for he not only arranges the whole of the British araniæ under greater and lesser divisions, but draws up a short and expressive specific character for each, which precedes his subsequent and more general description. Had this remarkable man imposed upon each species a single additional word, by which it could have been at once distinguished, — had he, in short given but a generic name to his groups, and a specific one to his species, — he would have been the first of nomenclators as he was of systematists; and the unbounded praise that has been so profusely lavished upon Linnæus for the simplicity of his distinctions, would have been more justly merited by Lister, inasmuch as the invention of precise systematic arrangement unquestionably belongs to the latter. Nor is this the only point in which Lister, so far as his researches extended, showed his superiority over the great Swede, who subsequently monopolised the applause of mankind. Lister looked to the habits and economy of these insects for the indications of their natural arrangement, all of which Linnæus, in his zeal for simplification, passed over; and thus fell infinitely below our countryman in his classification of the *Araniæ* no less than in his general arrangement of the testaceous animals. The greatest work of Lister, however, only appeared in 1685. It contains his general system, or synopsis, of conchology, and is enriched with no less than 1059 plates or figures of shells; among which several represent, with great accuracy, the internal structure of the animals themselves: most of these figures are so accurate, and all are so characteristic, that even to this day they are indispensable to the conchologist, and this remarkable volume forms one of the most valuable and standard works in this department of zoology.*

(13.) About this time natural history began to be pursued in England with greater zeal, and in a more philosophic spirit, than in any other part of Europe. No writer had appeared in France, since the days of Belon; nor had Italy contributed any thing to natural science, since the desultory yet curious observations of Boccone, the famous Sicilian botanist. On the other hand, Britain, which had been far behind in contributing to the early restoration of learning, seemed now to have suddenly sprung into life, and produced a constellation of

---

* Martin Lister. — (1.) Historiæ Animalium, Angliæ Tres Tractibus: unus de Araneis; alter de Cochleis tum terrestribus tum fluviatilibus; tertius de Cochleis Marinis. Londoni, 1678; small quarto. (2.) Historia sive Synopsis Methodica Conchyliorum. Folio. Londoni, 1685—1693. There is also a translation, by Lister, of Goedart's insects, published in 1685.

talent which left her without a rival. It is difficult
to trace, at this present time, the real causes which
led to this new and vigorous prosecution of science;
yet we are disposed to trace it, at least in part, to
the writings of the immortal Bacon, the effect of
whose sound philosophy first began to appear in the
land of his birth, where it disencumbered science
from the trammels of scholastic lore and ancient
tradition, teaching men to think for themselves,
and not to pin their faith upon the legends of
antiquity. Certain, however, it is, that simplicity
and perspicuity in writing upon the works of
nature first originated in this country, and that the
introduction of order and of system in the arrangement of his works, in this age of the world, entirely
originated from the great and united talents of
Lister, Ray, and Willughby, contemporaries of each
other, and alike directing their labours, though in
different departments, to one and the same object.
Of Lister we have already spoken, while the labours
of Ray and Willughby are so much interwoven,
that at first it appears difficult to decide which was
the most pre-eminent. Ray, with that candour and
simplicity which pervades all his writings, assigns
to his learned friend and patron the whole merit of
that ornithological arrangement which subsequent
writers have so erroneously given to himself[*]; nor
does it appear that Ray did more than augment the
" descriptions and histories" of his deceased friend's

---

[*] " Viewing his manuscripts after his (Willughby's) death,
I found the several animals in every kind, both birds, beasts,
fishes, and insects, *digested into a method of his own contriving.*"
Ray's Preface, p. 4.

Ornithology.* Of this work, and of its accomplished author, a short notice is all that we can give. Francis Willughby was a gentleman by birth and education, being connected with two noble families of that name, the Willughbys of Eresby in Lincolnshire, and the Wolds in Nottinghamshire: he was also allied by blood to the earls of Londonderry. We introduce these facts to show that science was even then not unknown among the aristocracy, and that young Willughby, although " endowed with excellent gifts and abilities both of body and mind, and blessed with a fair estate," which might have tempted him to seek pleasure and honour in the circle of the court, had yet such a love for learning, that " he was from his childhood addicted to study, and ever since he came to the use of reason so great a husband of his time, as not to let slip unoccupied the least fragment of it." That the greatest part if not all the original information contained in this admirable volume came solely from the pen of Willughby is also proved by the words of his pious editor †, who thus confesses that the additions made by himself were mere compilations from former writers.

---

* The Ornithology of Francis Willughby, of Middleton, in the County of Warwick, Esquire. In three books. By John Ray. 1678.

† " But because Mr. Willughby (though sparing neither pains nor cost) could not procure, and, consequently, did not describe, all sorts of birds to perfect the work, I have added the descriptions and histories of those that were wanting, out of Gesner, Aldrovandus, Bellonius, Marcgravius, Clusius, Hernandez, Bontius, Wormius, and Piso, disposing each kind, *as near as I could*, in its proper place."

Hence it becomes evident that the chief design of Willughby was only to admit into his history such birds as he had himself seen, or of whose existence there could be no doubt,—an admirable principle, in full accordance with the Baconian philosophy; and which, in this instance at least, establishes the superiority of his judgment over that of his editor. It was clearly with this view of acquiring original information that Willughby travelled in different parts of the Continent, " where he made so good progress in this work (his Ornithology) that few of our European animals, described by others, had escaped his view." And so ardent was his love of personal investigation, to the intent "that he might, as far as in him lay, perfect the history of animals, that he actually designed a voyage into the New World;" but the fiat of that beneficent Being, whose works he studied, and whose precepts he observed — for he was eminently pious — ordained otherwise: the hand of death arrested his bright career, and he died, in the year 1672, at the early age of thirty-seven. Willughby was the most accomplished zoologist of this or any other country; for all the honour that has been given to Ray, so far as concerns systematic zoology, belongs exclusively to him. He alone is the author of that system which both Ray and Linnæus took for their guide, which was not improved by the former or confessed by the latter. It has been customary for writers to represent Willughby more as a wealthy and intelligent amateur than as an original thinker; as the disciple and pupil of Ray in zoological pursuits rather than as his master and instructor. How far these opinions on their respective characters are supported by facts has been already shown. The system of Ray, in his Synopsis, is almost precisely a transcript from that of Willughby; and it is not one of the least beauties in the character of the survivor, that so far from wishing to appropriate to himself the laurels of his deceased patron, he seems particularly anxious to disclaim all pretensions to them. As the exposition of systematic details and of tables suits not with the nature of this sketch, we may at once pass from the patron to the *protégée*, and endeavour to form a just estimate of the real merits possessed by the third, though not perhaps the least, member of this zoological triumvirate. The life of Ray, unlike that of his friend, was protracted to a lengthened period; for he lived to the age of 77. The time, therefore, which he enjoyed for prosecuting his researches was nearly doubled: and hence he was enabled to expand them over a much wider field. Botany was his chief, if not his sole, study for the greatest part of his life; for he only began his work upon insects at the advanced age of seventy-five; although he had doubtless been collecting his materials for some time previously. In botany, and in no other science, was Ray the author of a system, for he confessedly adopted Willughby's, both in ornithology and ichthyology; while his arrangement of quadrupeds and of insects was doubtless derived from the same source. Indeed he himself informs us, that among the MSS. of his friend he found the histories of "*beasts* and *insects*," no less than of "*birds* and *fishes, digested into a method of his own.*" It belongs not to our present object to enquire into the merits of Ray's botanical system, which takes for its basis the old divisions of trees (*arbores*) and plants (*herbæ*). Suffice it to say, that although useful and even excellent when compared to former methods, this system has nothing very original in its structure, nor does it make the least approach to that masterly precision, which belongs to the arrangement of Linnæus. The merit of Ray, therefore, as a zoologist, must repose on his *Historia Insectorum**, published by Derham after his death. That we may not be accused of undervaluing the talents of this most amiable man, we shall quote the words of one who was well qualified to speak on the subject, and who was enthusiastic in his praise. " The descriptions given in the *Historia Insectorum*, especially considering the dark ages of this science in which they were written, are masterpieces of clearness and precision, and such as in general render it tolerably easy to ascertain the articles they belong to: but with respect to the arrangement and distribution of its materials, the work is in both these essential points unquestionably very far inferior to that of Linnæus; and, indeed, in some particulars, is not much superior to its predecessors. For, like them, it also incongruously blends the Linnæan class of *Vermes* with the genuine and natural one of insects!"† Having thus divested Ray of those inappropriate honours with which his memory has

---

* Historia Insectorum, autore Joanne Raio, &c. opus posthumum Jussu, Regiæ Societatis Londinensis editum. Londoni, 1710.

† Haworth, Review of Entomology.

been clothed, and which, had they been bestowed upon him when alive, he would have been the first to reject, we shall still find him a bright ornament to the age in which he lived, and in every way entitled to rank in the list of British worthies. True it is, that but for the patronage and protection of Willughby, he might probably have done little or nothing in science; and had he not been the editor of his patron's works, his name, *as a zoologist*, would have been far inferior to that of Lister, for he had neither the talents of the first, or the originality of the last: yet he laboured conjointly with both, and his name assumes a superiority from the variety of subjects he wrote upon, and from the number of works which bear his name, either as author or editor. Ray cannot be said to have possessed great genius, but he had sound judgment, great zeal, unwearied application;—a pious, amiable, and benevolent spirit, ever ready to acknowledge and to praise the labours of others, and do justice to their merits, even when he might have appropriated those merits to himself. But, above all, the name of Ray will ever be revered by the wise and the good, from the use he made of his extensive knowledge of nature. His " Wisdom of God manifested in the Works of the Creation" was the first attempt, we believe, ever made in the Christian era to confirm the truths of revealed religion by facts drawn from the natural world. Another of his works, " Persuasive to a Holy Life," shows us also how deeply his pure and pious spirit was imbued with those truths he taught to others. None but a philosopher could have written the first, none but a Christian the second. It is enough for the illustrious Ray that he united these characters in himself; nor should we, by investing him with fictitious qualifications, detract from the scientific fame of him who was his friend on earth, and who, we may humbly hope, is his companion in heaven!

(14.) From looking to the brightness cast upon the horizon of science by such names as Lister, Willughby, and Ray, we must now bestow a hasty glance on a few humbler men, who about this time aided, in different ways, the cause of natural history. Dr. Grew, better known as a botanist than as a zoologist, published, in folio*, and at the expense of the Royal Society, an account of the rarities in their museum; of which not a wreck is now left. It is worth remarking that in this extinct museum was the leg of a Dodo (mentioned at p. 60.): can this be the one now in the British Museum? or was it the companion? Grew's catalogue is a poor performance, although interesting to show how greatly natural history was at one time cultivated by the Royal Society. About this time, indeed, museums and collections were formed with much assiduity: the two most remarkable were those of Pettiver, a most zealous and indefatigable collector in all departments of nature; and whose museum was considered so valuable by his great but friendly rival, Sir Hans Sloane, that the latter eventually purchased it for the sum of 4000*l*. Pettiver was

---

* Museum Regalis Societatis, or a Catalogue and Description of the Natural and Artificial Rarities belonging to the Royal Society, and preserved at Gresham College, made by N. Grew, M.D. London, 1681.

a wealthy apothecary, and drew up a curious code of instructions for preserving animals and plants, which he gave to captains of ships, and other persons. These formulas, as may be supposed, were very rude, yet they contributed to fill the worthy apothecary's museum with a variety of new and curious objects; which he had engraved, without the least regard to order, and then published. Albin, also, who seems to have been a miniature painter, published in 1731 a quarto volume of 100 copper-plates, representing English lepidoptera in their different stages. As a work of art, this was, for the period, a very splendid undertaking; and, although devoid of any science, it must have materially advanced the cultivation of entomology. From the same hand originated, between 1731 and 1738, three volumes upon birds, and one upon British spiders; yet not of equal merit. We pass over the works of Bradley, Fermin, Klein, Knorr, Renard, Brown, and others of inferior note. But we may pause at the name of Sir Hans Sloane, then the most eminent patron of natural history in Britain; and holding the high professional station of court physician. A greater lover of natural history could not exist; for he expended a princely fortune in forming that museum and library which was ultimately purchased by the government, and made the foundation of the present national collection. Sloane, however, unlike the accomplished Willughby, was rather an amateur than a master, and his Natural History of Jamaica*, &c., although

---

* A Voyage to the Islands of Madeira, Barbadoes, Nevis, St. Christopher's, and Jamaica, with the Natural History, &c., by Sir Hans Sloane, Bart. 2 vols. folio. London, 1728.

it has never been superseded by a better, cannot be looked upon as having advanced either the precision or the arrangement of Zoology. Another of the great collectors of this period was Albertus Seba, of Amsterdam, who, like our Pettiver, was a wealthy apothecary. He collected all sorts of animals from all regions, and went to an enormous expense in publishing their figures and descriptions.* The engravings, for the most part, are very good, particularly those of the shells; but the descriptions are beneath criticism. Another splendid publication of this sort, in two folio volumes, was published by Catesby on the Natural History of Carolina†, which is even now very useful, from the plates being coloured, and tolerably accurate. The descriptions, likewise, are in general faithful, although destitute of any scientific merit.

(15.) Such were the ample materials existing in the year 1730, which the distinguished reformer of systematic Natural History, the great Sir Charles Linné (otherwise Linnæus), first began to model into shape; and which he ultimately condensed into the most simple, inviting, and luminous system the world had yet seen. The life of this extraordinary man is too well known, and has been too often written, to require any notice in this place; but his merits have been so extravagantly extolled by one party,

---

* Albertus Seba. Locupletissimi Rerum Naturalium Thesauri accurata Descriptio. Amsterd. 1734. 1765.

† The Natural History of Carolina, Florida, and the Bahama Islands. By Mark Catesby, F. R. S. Lond. 1731. 1743. With 220 plates. Another edition was edited by Edwards in 1771.

D

and so disparaged by another, that we may fairly enquire how far these conflicting opinions are founded in truth. That he was not the inventor of system, or of arrangement, even in his own age, is abundantly evident from the facts already stated: for the works of Lister and of Willughby were unquestionably his guides. Nor can he be said to have originated those large and comprehensive views in zoology, which had long ago been opened, like permanent lights in the firmament of science, by the immortal Aristotle. Great as were his talents and his genius, they were decidedly inferior to those of the Grecian philosopher. Neither had he at all times that accurate perception of affinities which can be traced both in the systems of Aristotle and of Willughby. His personal vanity, moreover, was excessive, — surpassing all bounds, and all instances upon record*; and this led him to do injustice towards some of those who were his contemporaries, no less than to Lister and Willughby, who were the real founders of scientific classification, and upon whose systems he framed his own. But, when we have said thus much, we have said all that can justly be charged against this illustrious naturalist. That he possessed great genius cannot be questioned, or he never could have conceived the herculean task of arranging all Nature; and without sound judgment and unwearied zeal he never could have accomplished his task. In his zoological works there is every indication of a powerful, comprehensive mind, while in his botanical writings

---

* Maton's Life of Linnæus, pp. 500. 561—563.

we trace the spirit of a philosopher. The services he rendered natural history, at the time he wrote, were immense; nor will they ever be forgotten. His unrivalled invention of nomenclature, which came from his hands, as it were, *perfect*, will remain of undiminished value so long as science exists; while the simplicity of those rules by which he arranged all the productions of nature then known, cannot be too closely imitated, however different may be the series in which these productions are disposed. He may be said to have created a language, peculiar to natural history, for the sole expression of the ideas pertaining to it:—a language which all, even his greatest opponents, are constrained to adopt, if they desire to be understood. Linnæus excelled in botany (for he loved it much more*), rather than zoology: but in both, his systems are confessedly artificial: the first will long be studied as a preliminary introduction to the natural system, but the latter, having served its purpose, above all others, in advancing the cultivation of zoology, has almost passed away in form, although not in spirit. His unrivalled invention of nomenclature, and the clear and lucid manner in which he arranged his materials, gave a facility to the cultivation of natural history, perfectly delightful; and

---

* This is quite evident, from observing the superior finish he bestowed upon his botanical works. And the following passage, in one of his letters to Ellis, places this partiality in a strong light:— " I care little about the larva of the iguana; but our mutual friend, Dr. Gardner, mentions some dried plants, destined for me. These I shall be very glad to have, whenever they arrive." — Linn. Corr. i. p. 178.

introduced a precision it had never before possessed. His object was to make known every natural production there discovered, in the most simple and concise terms; and to institute rules and forms of description, by which other objects, as they were progressively discovered, might be distinguished and registered in the same manner. In this he succeeded more completely than any who had then, or who has since, undertaken the same task. Unlike his great successor Cuvier, he knew the difference between a natural and an artificial system; he appreciated the value of the former, but he prosecuted his invention of the latter, because he saw it was more suited to the then state of science. That he possessed no inconsiderable knowledge of comparative anatomy, is abundantly evident; but he knew that the external characters of most animals were quite sufficient for the purpose of identifying them: and he wisely refrained from overburdening his definitions with unessential details and characters. Simplicity, in short, was his ruling passion, and it would be well for modern science if this principle had been imbibed by his successors. That he was ever anxious to improve his classification, to institute new divisions, modify his old genera, and make new ones, is attested by every succeeding edition of his *Systema Naturæ*, which he went on to improve until his death. These augmentations, however, were almost solely the result of personal knowledge. He possessed the spirit, the judgment, and the caution of Willughby, in rejecting all the vague and ill-defined species both of plants and of animals, mentioned by other writers and thus purged the science of a cumbrous and inextricable mass of "good-for-nothing lore," which confused, without instructing, the student. The amiable and gentle Ray, on the contrary, wanted the courage to do this: his own botanical works* are loaded with descriptions, even then obsolete, from the early writers, and he ingrafted the same useless lore, as he himself confesses, into the pure nervous descriptions of his master Willughby.

(16.) The publication of the *Systema Naturæ* gave to the study of Natural History a new form and a new life. Naturalists were astonished and delighted to see so much information condensed in so small a compass, and arranged in such luminous order. In those days, no other knowledge was sought for than the correct name of an animal or a plant,—whether it was known or unknown,—and what were its distinguishing characters. No wonder, therefore, that he who so admirably succeeded in communicating this information, insured immediate applause, and was suddenly raised to the rank of an oracle. His merits could be at once appreciated; no course of previous study was requisite to comprehend them, —no train of laborious investigation was essential to reveal their beauties. This was the true cause of the brilliant success experienced by Linnæus, and of the rapid adoption of his system. He rose into favour with his sovereign. Natural History counted kings and princes among her patrons. Linnæan Societies were formed in different parts of Europe; and the disciples of the great Swede travelled and

---

* Particularly in his Stirpium Europæarum extra Britannias Nascentium Sylloge,— a compilation from the works of Clusius, Bauhinius, F. Columna, &c.

collected in all the regions of the globe. The merits of that little band of British worthies, composed of Lister, Willughby, and Ray, whose writings brought about this sudden revolution in our science, seem to have been completely forgotten, in this general and exclusive homage paid to the great Northern Star. Yet England, as if determined to maintain her high character for original discovery, produced at this epoch one to whom even Linnæus himself was to bow. This genius was John Ellis, immortalised by the discovery of the true nature of the coralline animals, and by the masterly investigation he bestowed upon them. The value of this discovery is best stated in the words of Linnæus himself. In a private letter to Ellis he observes, " You have enriched our science by laying open a new submarine world to the admirers of nature *;" and " You have taken so lofty a rank in science, by your discovery concerning corallines, that no vicissitude in human affairs can obscure your reputation." No one more fully or more justly predicted the lasting fame of our celebrated countryman, whose discoveries were not confined, like that of Trembley on the polype, to a single genus, but comprehended a vast division of the animal kingdom.† Say what

---

* Linn. Corr. i. p. 164. 177.

† And yet, with these confessions, the unfortunate vanity of Linnæus prevented him from publicly confessing his own error regarding corals, and admitting to the full the splendid discovery of Ellis. " He has, consequently," as Sir James Smith truly observes, " fallen into half measures and ambiguities, which disgrace that part of his immortal Systema Naturæ, where these productions are described." — Linn. Corr. I.

we will, the best test of the merits of a writer, is the value which posterity attaches to his works; and, if we measure the researches of Ellis by this rule, we shall find, that, unlike systems claiming far higher pretensions, the two volumes of our illustrious countryman are now of as high an authority as they were on their first publication. His *Natural History of Corallines* *, now become scarce, was immediately translated into French, and we understand another edition has recently been published on the Continent. He was the author of no less than twenty-five papers in the Transactions of the Royal Society, and he was honoured by receiving the Copley medal for 1768 For some time previous to his death he had been gathering materials for a grand work on the zoophytes, and a considerable number of most admirable plates had already been executed, when this event took place, in 1776. These materials, however, were arranged by Dr. Solander, but only published † in 1786, under the auspices of Sir Joseph Banks; when both the author and the editor had gone to their last home. Ellis was also an accomplished

---

p. 80. He not only did this, but stoutly denied, to the last, in his own works, the discovery of Ellis. — See Maton's Life of Linn., p. 560.

* Ellis. (1.) Essay towards a Natural History of the Corallines found on the Coast of Great Britain and Ireland. By John Ellis, Esq. London, 1755. 1 vol. 4to. (2.) Letter to Dr. Linnæus on the Animal Nature of Zoophytes, called Corallina. London, 1768. 4to.

† Natural History of many curious and uncommon Zoophytes, collected from various Parts of the Globe. By Ellis and Solander. London, 1786. 1 vol. 4to

and an acute botanist, and his name must ever rank among the most endearing of those which add lustre to our science.

(17.) It is impossible, in the rapid survey we are now taking, to dwell upon all the names, much less to enumerate all the works, which now propagated the system of Linnæus, and gained fresh converts to the study of nature. For about fourteen years the illustrious Swede reigned, without a competitor, over the empire of zoology. But a formidable rival then arose, who divided with him the honours of supremacy. It will be necessary, however, before anticipating this part of our history, to notice a few writers, whose names occur between the years 1734 and 1754, or a period of about twenty years; during which time, nearly every thing that was published on *systematic* natural history emanated alone from Linnæus. Rumph (or, as he is more generally called, Rumphius), was a Dutch merchant resident for many years in Amboyna, during which time he investigated both the botany and conchology of that productive island, and he published the result in two separate works, still of much value.* There is a very interesting portrait of this venerable worthy, who nearly completed seventy years, prefixed to one of his works, representing him,—as he became in his latter days,—blind; yet still taking delight in examining his favourite shells by *touch*, when he could no longer do so by *sight*. The

---

* G. E. Rumphius. (1.) Thesaurus Imaginum Piscium Testaceorum, &c. Hagæ Comitum, 1739. folio. (2.) Cabinet d'Amboine, en Hollandois. Amst. 1705. 1 vol. folio. — *(Cuvier.)*

conchology of D'Argenville, which appeared in 1742*, although costly in its execution, has little claims to merit; the drawing of the figures, which constitutes its only value, renders it greatly inferior both to the volumes of Lister and of Rumphius. In 1758, conchology received another addition in the coloured plates of Regenfuss, published at Copenhagen: but the work was never completed, and the only volume that exists is so very rare, that we know but of two copies in this country.† A splendid addition was made to illustrated entomology in 1746, by the coloured figures of Rœsel‡; and here also we may notice the valuable collection of figures by our countryman Edwards; whose works, although terminated at a time when most writers arranged their materials according to the Linnæan method, were commenced in 1743, and belong to the illustrative, more than to the scientific class, of zoological publications. Edwards was the friend of Sir Hans Sloane, and for many years filled the office of librarian to the College of Physicians. He has no pretensions to scientific talent, or to original research; yet it is an extraordinary fact, that, destitute of such

---

* D'Argenville. L'Histoire Naturelle éclaircie dans une de ses principales Parties, la Conchylcologie. Paris, 1742. Another edition appeared in 1757; and a third, augmented by Favanne, in 2 vols. 4to, in 1780.

† Regenfuss. Choix de Coquillages et de Crustacés. Copenhagen, 1758. Folio.

‡ A. J. Rœsel. Der Monatlich, Herausgegebenen Insecten Blustigung; or, a Monthly Publication on the Amusements of Insects. Nurenberg, 1746—1761. 4 vols., small 4to. The fourth volume is a Supplement by Kleeman.

qualifications, his works are assuredly the most valuable, on general ornithology, that have ever appeared in England. This arose from his being the first who figured and described a vast number of birds, then new to naturalists, who consequently refer to him as the original authority for all such species. The figures of Edwards were copied and recopied by nearly all succeeding writers, up to the year 1820, when we ventured, in the *Zoological Illustrations*, to introduce a new style of delineation; and to substitute original figures for those which were then copied into nearly all the popular compilations. Edwards was remarkably exact in his descriptions, and sufficiently so in his figures, so that no zoological library, especially one for reference, should be without his volumes.* In the year 1744, the famous discovery was made by Trembley, a native of Geneva, of the reproductive powers of the freshwater polype.† The developement of this wonderful fact entitles his name to a high station in the records of analytical research; although, in its general effect, this discovery exercises far less influence on zoological arrangement

---

* Edwards. (1.) The Natural History of uncommon Birds, and of some other rare and undescribed Animals. By George Edwards, Library Keeper to the Royal College of Physicians. In 4 vols. 4to. 1743—1750. Edwards is erroneously termed by Cuvier, *Peintre Anglais*. (2.) Gleanings of Natural History, exhibiting Figures of Quadrupeds, Birds, Insects, Plants, &c. By George Edwards, F.R.S. and F.A.S. In 3 vols. 4to. 1758—1764.

† Trembley. Mémoires pour servir à l'Histoire des Polypes d'eau douce, à bras en forme de cornes. Leyden, 1744. 1 vol. 4to.

than that of Ellis. Ichthyology had received no additions since the time of the illustrious Willughby, but in 1754, Gronovius, a wealthy collector of Leyden, published an account of the fish contained in his museum.* This work, now become very rare, we do not possess. It is evident, however from a subsequent publication†, that Gronovius received the Linnæan system with a strong and even an absurd prejudice. He quotes the works of its author, it is true, but he neither adopts his specific characters, or his nomenclature, preferring to designate his subjects after the old method, rather than by a positive name. So strongly, indeed, does he seem prejudiced on this point, that he carefully excludes the specific names of Linnæus from his entire work! Gronovius, however, gives the characters of several genera, not to be found in the *Systema Naturæ*, and so far his opposition to Linnæus was perfectly justifiable; while his descriptions and characters are copious and excellent.

(18.) But it is time for us to notice a new school of naturalists, which arose in France even before the publication of the first edition of Linnæus's system, and which was brought about by the celebrated Reaumur's Memoirs towards a History of Insects ‡; the first volume of which was published at Paris, in

---

* Laur. Theod. Gronovius. Museum Ichthyologicum. vol. in fol. Leyden, 1754. — (Cuvier.)

† Zoophylasii Gronoviani, exhibens Animalia quadrupeda, Amphibia, atque Pisces, &c. Lugd. Batav. 1763. 1 vol. fol.

‡ Mémoires pour servir à l'Histoire des Insectes, par M. de Reaumur Paris, 1734—1742. Six vols. 4to.

1734. There can be no doubt that it was this valuable and beautifully written work, full of interesting facts, detailed in popular and elegant language, that first induced Buffon to adopt a similar style, and to clothe natural history in such a dress that it should interest the world. That he completely succeeded in so doing, by those graces of composition and those charms of eloquence which he possessed, is notorious to all. These qualifications were his own, but they would have been altogether useless, at least in this undertaking, but for the sound information, the knowledge, and the experience of his friend and fellow labourer. Daubenton, who supplied the eloquent biographer of the animal kingdom with that solid information he did not possess, and without which comparatively he could have done nothing. It is unreasonable to expect that a man like Buffon should excel in such opposite qualities as rigid and laborious research, cautious deduction, and flowery eloquence. Upon the two first is built every thing valuable in pure science; while the latter, however desirable, is merely ornamental;—it may captivate the world, but it is rather detrimental than otherwise to the advancement of sound knowledge, and the calm investigation of truth. Hence it led the vivid and excursive fancy of Buffon into wild and fanciful theories, positive assertions, and palpable blunders. And these errors, although clothed with all the charms of eloquence, faded away—like the mists of a summer morn—before the rays of truth. That the writings of this celebrated man promoted, indirectly, the extension and the advancement of natural history, is beyond all doubt; but it may safely be affirmed, that its science, or its philosophy, derived little or no direct benefit from his splendid compositions. His character and his writings were in complete accordance with those of the nation to which he belonged; and his immediate popularity was the consequent result. From that time the cultivators of zoology were divided into two schools; one party following the systematic investigation inculcated by Linnæus, the other ranged themselves under the banners of Buffon, and gathered the flowers, without probing for the honey.

(19.) The disciples of Linnæus, whose proceedings we shall first briefly sketch, followed, with little deviation, the line of enquiry and the plan of arranging their discoveries, pursued by their great master. Artedi, who was among the earliest and most eminent disciples of this school, studied fish: and his Ichthyology is one of the most valuable treatises of those animals we even now possess. By the recommendation of Linnæus, the wealthy Seba intended to have engaged him in describing and arranging the ichthyological portion of his voluminous work, already alluded to: but by the sudden and premature death of young Artedi, occasioned by his falling into one of the canals on returning at night to his lodgings, this project was defeated. Linnæus, who edited the works of his friend and pupil, prefixed to the volume an interesting Life of its author, which will be perused with pleasure.* The curious reader will find in the first part an

---

* P. Artedi. Ichthyologia, sive Opera omnia de Piscibus. Lug. Bat. 1738. 1 vol. 8vo.

admirable and erudite chronology of all the writers upon ichthyology, from the most remote records, up to the time of Willughby and Ray. It is singular that no specific names appear in this volume, although edited by Linnæus: and, while ample descriptions are given of such as were new species, little is stated in regard to others, beyond innumerable references to ancient writers. Artedi fell into the prevalent notion of considering whales, and all the cetaceous quadrupeds, as true fish: but with this exception he so far surpassed all his predecessors in clearness of arrangement, and in the extent of his materials, that he deserves to be considered the father of systematic ichthyology. Artedi was followed in this department by Gronovius, whose name we have excluded from this school, rather on account of his strange rejection of the Linnæan nomenclature, than from a departure from that mode of arrangement which originated with the great Swede. Sulzer was the first who adopted the Linnæan entomology, for in 1761 he published a work with coloured plates, expressly to illustrate this system of insects*; and this was followed, fifteen years after, by others, having the same object.

(20.) Entomology now began to be pursued with much avidity on the Continent, but more especially in Germany; where, to this day, it has continued to flourish more than in any other part of

---

* J. H. Sulzer. (1.) Die Keunzeichen der Insecten, &c.; or, The Characters of Insects, according to Linné. Zurich, 1761. 1 vol. 4to. (2.) Abgekürzte geschichte der Insecten, &c.; or, The abridged History of Insects, according to Linné. Winterthur, 1776. 2 vols. 4to.

Europe. No better evidence of these facts can be adduced than the rapid increase of new works in this department towards the middle of the last century. Entomology, in the following year, was enriched with the most inimitable delineations of insects which this or any age has produced; and which form the plates to that beautiful work by Sepp (in Dutch) on the insects of the Low Countries.* This publication came out in numbers, and perfect sets are now exceedingly rare: those portions we possess relate exclusively to the Lepidoptera, each species being delineated, in all its several transformations, from the egg to the perfect insect: the drawing of the subjects is chaste, elegant, and cannot be excelled for accuracy; while the style of engraving is admirably suited to express all the softness of the original drawings. These plates, in fact, have never been equalled, far less excelled, by any of the most celebrated in modern times. Sepp undertook, in like manner, to figure all the birds of his native country, but his talents were quite unsuited to this department; and his figures have all the stiffness and roughness of badly preserved dried specimens. The works of Sepp, who is the Van Huysum of our science, are more illustrative than scientific, while that of Scopoli, on the entomology of Carniola†, which soon followed, is purely descriptive: he does not, however, implicitly follow Linnæus in the names of

---

* Sepp. Beschouwing der Wonderen Gods in de Minstgeachte Schepzelen of Nederlandsche Insecten. Amsterdam, 1762, &c. 3 vols. 4to.

† J. A. Scopoli. Entomologia Carniolica, exhibens Insecta Carnioliæ indigena. Vindobonæ, 1763. 1 vol. 8vo.

his orders, and he proposes several new genera: but he writes in the spirit of the *Systema Naturæ*, and the excellence of his descriptions shows he was an accurate observer and a really good naturalist. Scopoli was Botanical Professor at Pavia. With that moral courage which bespeaks an honest and a good heart, he had the " temerity" to expose the disgraceful thefts made by Spalenzani of objects from the public museum. The interest of the accused, however, supported him; and although the proofs adduced were unanswerable, the remainder of Scopoli's life was rendered miserable by the persecution of Spalenzani's friends. He was subsequently the author of three other works*, and he is stated to have published some plates, illustrating his Entomologia Carniolica, but which we have never met with in any library. Three years afterwards, the industrious Schœffer, of Ratisbon, began to publish his voluminous and expensive works, chiefly upon the insects of his native province: they are now valuable only for the numerous coloured figures, of poor execution as works of art, yet very useful for reference. He was not altogether a disciple of Linnæus, for he endeavoured to set up a system of his own, of which he published the Elements †,

---

* Scopoli. (1.) Introductio ad Historiam Naturalem. Pragæ, 1777. 1 vol. 8vo. (2.) Anni Historico Naturales. Lipsiæ, 1768—1772. 1 vol. 8vo. (3.) Deliciæ Floræ et Faunæ insubricæ. Ticini, 1786—1788. 1 vol. folio.

† J. C. Schœffer. (1.) Elementa Entomologica. Regensburg, 1766. 1 vol. 4to. In Latin and German. (2.) Icones Insectorum circà Ratisbonam indigenorum. Regens. 1769. 3 vols. 4to.

containing many new genera, not to be found in Linnæus; yet the characters are short and unsatisfactory; and, strange to say, he nowhere uses specific names. Schæffer was a clergyman of Ratisbon, and lived to the age of seventy-two; but, although industrious, his abilities were very moderate.

(21.) While entomology was thus advanced by describers and painters, the disciples of Linnæus were returning from their travels, and pouring into the lap of their master the innumerable novelties they had discovered in distant regions. Hasselquist, who had been travelling in the East, published his narrative in 1757. Osbeck returned from China, loaded with its plants and animals. Forskal* was no less zealous and successful in investigating the little known tracts of Egypt and the shores of the Red Sea: and Sparmann† travelled both to Southern Africa and China. The treasures collected by these enterprising and accomplished travellers went to augment the accumulating materials of Linnæus, and rapidly swelled the bulk of each succeeding edition of his *Systema Naturæ*. It is really surprising to witness with what rapidity this celebrated man could arrange and incorporate materials so numerous and so varied, as they came pouring in upon him from all quarters, and which, to ordinary men, would have been perfectly overwhelming. While Zoology was thus proceeding with rapid strides upon

---

* Forskal, P. Descriptiones Animalium, &c., quæ in Itinere Orientali observavit. 1775. 4to.—Icones Rerum Naturalium quas in itinere Orientali depingi curavit. 1776. 4to.

† Sparmann, A. Museum Carlsonianum. Holmiæ, 1786—1789. Four parts, forming 2 vols. small folio.

the Continent, it seems to have made no great progress among us, were we to judge from the paucity of works then published in Britain. Edwards, it is true, was going on with his excellent *History of Birds;* and Borlase had done something to illustrate the natural history of Cornwall; but no new work of any moment appeared in England between the years 1755 and 1766; when our accomplished countryman Pennant gave that impetus to the science, which it seems to have required. Pennant was a scholar and a gentleman, possessing great and varied acquirements. He was versed in classic and in historical learning; and passionately attached to the natural history of his own country. His works, in all these departments, are numerous; but he is chiefly known among us as the first who treated the natural history of Britain in a popular and interesting style.* He followed the system of Linnæus, except in that strange and unnatural arrangement of the primary orders of birds, which he fell into, and which was the more inexcusable, after the writings of Willughby. There are no novelties of arrangement in the works of Pennant, and no original research, beyond the accession of new species; but he contrived to give great interest to his descriptions, by enriching them from the stores of his classic and antiquarian knowledge. Hence he enjoyed great

---

* Thomas Pennant. British Zoology, 1 vol. folio. London, 1766. Ditto, in 4to. Ditto, in 8vo., 4 vols. 1812. — Synopsis of Quadrupeds. Chester, 1771. 1 vol. 8vo. — History of Quadrupeds. London, 1793. — Genera of Birds. London, 1781. 1 vol. 4to. — Indian Zoology. London, 1790. 1 vol. 4to. — Arctic Zoology. London, 1792. 2 vols. 4to.

popularity; and his works may be perused, even now, with pleasure and advantage. Although he must have greatly contributed to extend a taste for these pursuits, yet his example assisted, without doubt, to throw upon our succeeding writers those fetters of implicit obedience to the authority of Linnæus, which every fresh example more firmly riveted; until at length it was deemed a sort of heresy to propose a new division, or to name a new genus. Pennant for many years held a constant correspondence with the ingenious and amiable White of Selborne; who, though not a professed writer upon systematic natural history, contributed very much to the information of Pennant, and whose popular and interesting letters have recently been published by so many different editors. White, in short, was one of those very few who then devoted his attention to the observance of nature, without making any attempt to generalise the facts so acquired. Natural history, to such observers, is but a mere amusement, fascinating indeed, and even useful, but totally disconnected with the objects of philosophic science. Entomology, which had been so much advanced on the Continent by the figures of Rœsel, Sulzer, Sepp, and Schæffer, and by the scientific volume of Scopoli, now began to make some progress in England; more indeed by the admirable figures of Moses Harris, than by the descriptions which accompanied them in the three volumes of Drury's Exotic Insects *; the first of which appeared

---

* Illustrations of Natural History. By D. Drury. London, 1770—1772. 3 vols. 4to.

in 1770. Drury was a wealthy jeweller, and expended large sums in sending out practical collectors to all parts of the world, to enrich his cabinet with new insects. It is to one of these, Mr. Henry Smeathman, that we are indebted for an elaborate and most interesting account of those wonderful insects, generally termed white ants; this remarkable discovery was published in the Philosophical Transactions, and subsequently translated into French.* This is unquestionably one of the most valuable discoveries in the natural history of insects ever made; yielding only to that of Huber's on the bees.

(22.) Nearly at the same time that the first volume of Drury's Insects was published in England, the great work of Martini†, on General Conchology, made its appearance in Germany. This bold and costly undertaking at once shows how great a taste for shells then existed; for it extended, with the continuation by Martini, to no less than eleven quarto volumes; and, notwithstanding the poorness of its figures, it still continues to be one of the standard authorities for reference in this department: the arrangement, however, is defective, and it possesses none of the judgment or the correct views of Lister. In 1773, another addition to the already numerous collections of entomological figures was made by Benjamin Wilks, who published 124 plates of English moths and butterflies.

---

* Mémoire pour servir à l'Histoire de quelques Insectes connus sous les Noms de Termis ou Fourmis Blanches. Par M. H. Smeathman. Ouvrage rédigé en François par M. Cyrille Rigaud. Paris, 1786. 8vo.

† Martini und Chemnitz. Neue Systematisches Conchilien Cabinett. Nurnb. 1769—1800.

(23.) It deserves notice, that, notwithstanding the numerous works upon entomology that appeared between 1734 and 1773, they were all, excepting Scopoli's, more or less illustrative; that is, intended to delineate insects, rather than to describe them. The letterpress, in fact, was subordinate to the plates; so that all that the science gained was an immense accession of new species, requiring the institution of new genera, and new divisions for their reception in scientific arrangement. This task was undertaken, in 1775, by a distinguished disciple of Linnæus, the celebrated Fabricius, who in that year commenced the publication of his voluminous works*, which subsequently extended to nearly twenty octavo volumes. Fabricius, although in one sense the founder of an entomological system, was nevertheless a disciple of that purely systematic school, of which we are now tracing the progress. Had he been content to have increased the genera of his original instructor, to suit the vast additions that had now been made to the knowledge of species and groups, his fame would have been equally brilliant and more lasting: but, like very many of those who went before, and who came after him,

---

* Jo. Christ. Fabricius. Systema Entomologiæ. Lipsiæ, 1775. 1 vol. — Species Insectorum. Hamb. 1781. — Supplementum Entomologiæ Systematicæ. Hafniæ, 1798. — Genera Insectorum. Chilonii. — Mantissa Insectorum. Hafniæ, 1787. — Philosophia Entomologiæ. Hamb. 1778. — Entomologia Systematica, emendata et aucta. Hafniæ, 1792—1796. 5 vols. — Supplementum Entomologiæ Systematicæ. Hafniæ, 1798. — Systema Eleutheratorum. Kiliæ, 1801. 2 vols. — Systema Piezatorum. 1804. — Systema Antiliatorum. Bruns. 1805.

he chose to affect entire novelty; he made alterations and innovations in matters where none were called for; and he built his arrangement upon characters which, when taken by themselves, are not only extremely difficult of detection, but also very artificial. He devised new names for the orders, and he founded his generic characters entirely on the parts of the mouth, excluding all others of external structure. It seems difficult to account for the great popularity which Fabricius at one time enjoyed on the Continent, and even in England; seeing that, although his zeal and industry were unwearied, his principles of classification were troublesome and complicated, and his ideas on the philosophy of his science crude and superficial. But this popularity entirely arose from his having no competitor in systematic entomology. Linnæus, having defined his orders, and indicated the chief generic groups, seems to have almost relinquished further improvements in this branch of his studies, and to have tacitly resigned entomology into the hands of his pupil. The consequence was, that the entomologists of the day, continually discovering new species, and finding the Linnæan genera totally inadequate to contain such accumulating novelties, had no alternative but to adopt the system of Fabricius, at least so far as his genera were concerned, for very few were disposed to relinquish the Linnæan names of the orders; and Fabricius, in some of his subsequent works, was induced to bring in the external characters of his genera, in addition to those taken from their oral organs. Fabricius lived long, and wrote much; so that he may be said to have presided over systematic entomology for nearly thirty-two years; that is, from 1775 to the beginning of the present century: he lived, however, to see the rapid declension of his system, before the rising star of the celebrated Latreille.

(24.) Those who still adhered to the entomological arrangement of Linnæus, were Thunberg, one of his most eminent disciples, who travelled in China, Japan, and Southern Africa, and ultimately filled the botanical chair of Upsal; Müller*, who wrote a valuable *Fauna* of the animals of Denmark and Norway; Forster†, the companion of Captain Cook, who has left us a *Century of Insects;* and Villers, who‡, even so late as 1789, made a vain and retrograde movement in the science, by reducing all the genera of Fabricius and of others to the Linnæan standard. We may here mention the excellent work of Schrank §, who systematically investigated and described the insects of Austria.

---

* O. F. Müller. Zoologiæ Danicæ Prodromus. Hafniæ, 1776. 1 vol. 8vo. Also, Fauna Insectorum Fridrichsdalina. Hafniæ, 1764. 8vo.

† J. R. Forster. Novæ Species Insectorum. Centuria 1 Londini, 1771. 8vo. — A Catalogue of British Insects. Warrington, 1770. 8vo.

‡ Villers. Car. Linnæi Entomologia, Faunæ Suecicæ Descriptionibus aucta; D. D. Scopoli, Geoffroy, De Geer, Fabricii, Schrank, &c., Speciebus vel in Systemate non enumeratis, vel nuperrime detectis, vel Speciebus Galliæ Australis locupletata, Generum Specierumque rariorum Iconibus ornata; curante et augente Carol. de Villers. Lugduni, 1789. 3 vols. 8vo.

§ F. Schrank. Enumeratio Insectorum Austriæ indigenorum. Aug. Vind. 1781. 8vo.

Of illustrated works belonging to this school, there are two of great merit. Fabricius found a powerful supporter in the celebrated French entomologist Olivier*, who, after travelling extensively in Turkey, Egypt, and Persia, returned, and commenced his great work upon coleopterous insects; the richest in figures and in description we yet possess. Rœmer†, also, has illustrated the genera of Fabricius with remarkably good figures, drawn with a boldness rarely seen in those of other artists. But the most splendid work of this description relative to British insects, which appeared in this era, is the *Aurelian* of Moses Harris‡, the executor of Drury's figures, already mentioned, and whose beautiful plates far exceed those of Albin, Wilks, or Donovan, on the same subject. Harris cannot be regarded as a scientific entomologist, yet it is curious to trace the perception he had of natural arrangement. He was the first, in fact, who distributed all the British *Diurnal Lepidoptera* into those genera termed modern, long before those who have the credit of so doing were born. The excessive rarity of the little tract which substantiates this fact, so honourable to our countryman, is no doubt the reason why it has never been noticed§, a

---

* Olivier. Entomologie, ou Histoire Naturelle des Insectes. Paris, 1789—1808. 5 vols. 4to.

† J. J. Rœmer. Genera Insectorum Linnæi et Fabricii Iconibus illustrata. Vit. Helv. 1789. 4to.

‡ Moses Harris. The Aurelian; or, Natural History of English Insects, namely, Moths and Butterflies. London, 1778. folio.

§ Moses Harris. An Essay preceding a Supplement to the

publication upon the *Lepidoptera*, inferior, indeed, to that of Harris in execution, but much more comprehensive, was commenced, in 1779, by Cramer *, and terminated by Stoll; comprising, upon 442 plates, the largest collection we possess of coloured figures of the exotic *Lepidoptera*: the fifth volume is entirely by Stoll †, and is enriched with numerous representations of larvæ and pupæ, drawn in Surinam. Although these figures are indifferently drawn and coarsely coloured, yet they are nevertheless sufficiently accurate to render this work of much value.

(25.) Leaving entomology for the present, let us trace the progress of the Linnæan school in other departments. The work upon quadrupeds, by Schreber, commenced in 1775, is of little value; most of the plates being copies, and inaccurately coloured. But in 1776, the most celebrated of all the disciples of Linnæus, Professor Pallas, began to publish his various essays and dissertations on almost every branch of zoology.‡ The commencement of his

---

Aurelian. London, no date: but at the corner of the 2d and 6th plates is inscribed, " M. Harris del. et sculp. Oct. 20. 1767." — An Exposition of English Insects. London, 1782. 1 vol. 4to. — The English Lepidoptera; or, the Aurelian's Pocket Companion, &c. London, 1775. 8vo.

* P. Cramer. Papillons Exotiques des Trois Parties du Monde; l'Asie, l'Afrique, et l'Amérique. Par M. Pierre Cramer. Amsterdam, 1779, &c. 4 vols. 4to.

† C. Stoll. Supplément à l'Ouvrage du M. Cramer. Amsterdam, 1791. 1 vol. 4to.

‡ Pallas. Miscellanea Zoologica. 1776. 1 vol. 4to. — Spicilegia Zoologica. Berl. 1767—1780. 1 vol. 4to. — Novæ Species Quadrupedum e Glirium Ordine. Erlang

career gave little promise of his subsequent proficiency; for he rashly and ignorantly entered the lists against the celebrated Ellis, and maintained that corallines were plants! * Pallas was engaged by the court of Petersburgh for many years: he travelled

---

1778. 1 vol. 4to. — Icones Insectorum præsertim Rossiæ Siberiæque peculiarium. Erlang. 1781. 2 Nos.

* Ellis thus writes to Linnæus: — " There is now printing, in Holland, a book on Zoophytes, by Dr. Pallas of Berlin, who was two years in England. This gentleman, I find, has treated both you and me with a freedom unbecoming so young a man. I find Pallas has used me with so much ill-nature, because I exposed the absurdities of (his friend) Baster's doctrines and experiments, in our Phil. Trans." (Linn. Corr. i. p. 186.) Again : — " Dr. Pallas, in his article of Corallines (vide Pallas, Zoophytes, p. 418.), depending on Count Marsigli's chemical analysis of them, considers them as vegetables. But if we observe how Pallas has confounded the calcareous crust of corallines with the farinaceous covering of vegetables, it will be no longer a matter of surprise: for had he put the true corallines into an acid menstrum, and the *Fucus pavonius*, which he calls *Corallina* pavonia (Pall. Zoophy. 419.), and the *Lichen* fruticulosus, which he calls *Corallina* terrestris (vide p. 427.), he would have found that the true corallines would ferment strongly, while the *Fucus* and *Lichen* would not be in the least affected." (Linn. Corr. i. p. 198.) The high praise bestowed upon Pallas in the *Règne Animal*, and the slight notice taken of Ellis in the same work, is the occasion of this note. That Pallas published a great deal more than Ellis, is very true, because the one was by profession a naturalist, in the service of Russia; while the other held a high and responsible appointment under government, and could only pursue natural history at his leisure. But talents are not to be esteemed in this way: and the subsequent confession, *by Pallas himself*, of his errors (Linn. Corr. i. p. 227.), places the relative powers of these observers in their true light.

extensively both in Europe and Asia; and described with more than usual accuracy the animals he met with; he was also a very good comparative anatomist, and having no other profession to distract his attention, and being blessed with a long life, he had time to acquire considerable knowledge in all departments of nature, not excepting botany and mineralogy. He was more especially engaged by Catherine II. to travel through the Asiatic provinces of Russia, with a view to investigate their natural productions. His travels, published at the expense of his munificent patron, were translated into French, and subsequently into English. Pallas was undoubtedly the most accomplished zoologist of the Linnæan school, and, if he was not the author of any striking or important discovery, he accomplished more, in other ways, than any one of the era in which he lived. Two works upon systematic conchology appeared in 1779, by Schrœter* and Born†, illustrated by figures; those of the latter are very well drawn, and delicately coloured, but those of Schrœter, in this and his subsequent publications, are indifferent, even for this period. Merram ‡

---

* Schrœter. A Treatise on River Shells (in German). Halle, 1779. 4to. — An Introduction to the Linnæan System of Conchology (in German). Halle, 1783—1786. 3 vols. 8vo. — An Account of the internal Structure of Sea Shells, &c. (in German). Frankfort, 1783. 1 vol. 4to.

† Born. Testacea Musei Cæsarei Vindobonensis. Vienna, 1780. Folio.

‡ Merram, B. Avium rariorum et minus cognitatum Icones et Descrip. Leipzig, 1786. 1 vol. 4to. — Materials for a Natural History of Reptiles (in German). Duisbourg, 1790. 2 parts, 4to.

in 1789, published the descriptions and figures of some new birds; and subsequently, in 1790, he took up the examination of the much neglected class of reptiles, intending to treat upon them in detail, but, unfortunately, the work only reached to the second number: this was the more to be regretted, since Merram had evidently paid great attention to these animals. Hermann, the professor of Strasburg in 1783, deserves to be particularly mentioned, as much for his systematic descriptions*, published after his death, as for his very curious and valuable work on the affinities of animals, wherein he brings into comparison individuals of different orders, resembling each other. These tables are well worth the perusal, and even study, of the philosophic zoologist; for though Hermann was perpetually confounding the two relations of affinity and analogy, yet we can here trace the faint germ of those enlarged views on the natural system, which, after a lapse of many years, were to be so much expanded. His son inherited the taste of his father, and published a work on apterous animals; but this we have not seen.

(26.) Ichthyology, one of the first departments of natural history which engaged the attention of the writers of the sixteenth century, had received but few additions since the time that Linnæus began his splendid career. Dr. Garden, one of his most valuable and learned correspondents †, had supplied him

---

* J. Hermann. Observationes Zoologicæ Posthumæ. Strasburg, 1804. 1 vol. 4to. — Tabula Affinitatum Animalium. Strasburg, 1783. 1 vol. 4to.

† See Linn. Corr.

with many new species from America; and the coloured figures of Catesby had made known several others, peculiar to the coasts and rivers of Carolina. Müller, likewise, had touched upon the species of Denmark, and a few had been described by Forskal from the Red Sea; but these additions to ichthyology were very insignificant, when compared to those which other branches of zoology had received; while the want of good figures, even of the species already known, left the knowledge of these animals in a very backward state. The appearance, therefore, of the famous work of Bloch, who began the publication of his great undertaking in 1785, must have been hailed with pleasure. Bloch was a Jewish physician, settled in Berlin; and his *Ichthyology*, in twelve folio parts, containing no less than 452 coloured plates, was such an undertaking as no one would have courage to prosecute in these days, unless with the determination of submitting to a large pecuniary sacrifice. It is, without doubt, the most complete work, in regard to figures, that has ever been published: for although the subject was treated of subsequently by La Cépède in greater detail, and with a considerable addition of species, the figures in the French work are small, uncoloured, and not altogether remarkable for accuracy.* Bloch, although, like Fabricius, the author of a system, followed the systematic style of arrangement pursued by Linnæus; and both his characters and his descriptions are excellent. One only regrets that a work so essential to

---

* M. E. Bloch. Ichthyologie, ou Histoire Naturelle générale et particulière des Poissons. 12 parties, folio. Berlin, 1785—1796.

every ichthyologist, is, of necessity, so expensive. There is, however, a smaller edition in octavo *, in six parts or volumes, with 216 coloured plates, which, so far as it extends, is equally useful with the folio edition. Bloch has the great excellency of describing such species only as he had himself seen; a rare quality in the writers of this period, when compilations began to be made in the shape of general systems, which almost brought us back to the age of Gesner and Aldrovandus. Schneider†, however, who published what he termed the system of his friend, after his death, added separately in two volumes the species described by other authors: but this work we have not seen. Bloch was also the author of a volume on the intestinal worms.‡ His continuator, Professor Schneider, was also attached to the study of the *Amphibia*, upon which he wrote some dissertations of great merit §; particularly on the Tortoises, a tribe which was again illustrated in 1792 by the coloured plates of Schœpf.||

(27.) Ornithology, as will subsequently appear had been much attended to by the disciples of Buffon;

---

* Idem, en six parties, avec 216 planches. Berlin, 1796 8vo.

† J. G. Schneider. Systema Ichthyologiæ de Bloch. 2 vols. 8vo. avec 110 fig. Berlin, 1801. (Cuvier.)

‡ Traité sur la Génération des Vers Intestines. Berlin, 1782. 4to.

§ Schneider. General Natural History of the Tortoises (in German). Leipzig, 1783. 8vo. — Amphibiorum Physiologiæ Specim. 1 et 2. Zullichau, 1797. 4to.

|| Schœpf. Historia Amphibiorum, Naturalis et Litteraria. Fas. 1. et 2. Jena, 1799 and 1801. 8vo. — Historia Testudinum Iconibus illustrata. Erlang. 1792. 4to.

but the vast accession of new species, which now required systematic arrangement, pointed out the necessity of a general work on this subject. Linnæus had now ceased from his labours, for he closed his bright career in 1778; but his system was still paramount in this and most other countries ; and in 1782, our celebrated countryman, Dr. Latham, adopted it in his *General Synopsis of Birds*, save only in the primary divisions. This great and laborious undertaking was brought to a close in 1790, and it remained, for many years, the best descriptive catalogue of birds extant. Several new genera were proposed ; but it was not the practice, at this time, to pay much attention to the minutiæ of structure. It was thought sufficient, for instance, for the purposes of arrangement, to refer all flat-billed perching birds to the genus *Muscicapa*, and that of *Sylvia* contained all those with slender straight bills. As no effort was made to improve the definitions of the Linnæan genera, or to restrict them within due limits, it necessarily followed that the same species was not unfrequently described two or even three times, under as many different names, and in different genera; while the desire of the author to include all the species of birds then known, induced him to transcribe from other authors the accounts of such as he had not seen himself; and to introduce as distinct species numberless others, whose existence

---

* Dr. Latham. (1.) A General Synopsis of Birds. By Dr. J. Latham. 3 vols. and 2 Supplements. 4to. London, 1782, &c. (2.) Index Ornithologicus. London, 1790. 2 vols. 4to. — (3.) A General History of Birds. Winchester, 1821—1824. 10 vols. 4to.

rested only upon the faith of drawings in the hands of his friends. From these circumstances, numberless errors inevitably resulted: and, with all our respect for the venerable author, we are compelled to confirm the judgment already passed upon this work by Cuvier.* Nevertheless it must be admitted that a large number of really new and most interesting birds were now, for the first time, sufficiently well described; and that, at the period when they were published, both the *General Synopsis* and the *Index Ornithologicus* were useful and even valuable publications. They accomplished, "in their generation," the object for which they were intended, — they advanced science; while their very imperfections brought about that revolution in our mode of investigation, which has now rendered them of little service. We should have wished, for the reputation of the first writer whose works we studied, that the *History of Birds* had never appeared; since it is merely an enlargement of the *Synopsis*, presenting us, in the year 1820, with the systematic views which were prevalent in 1782; a system, in short, which, having served its turn, is now only a matter of history. We feel pained at being called upon to criticise the works of authors who are now living, for it will surprise most of our readers when they are told that the amiable and venerable author of the *Synopsis* is now enjoying a vigorous old age, having outlived, if report speaks true, ninety-four winters. Should these remarks ever meet his eye, we pray him to pardon their freedom; and we entreat him to re-

---

* Règne Animal, vol. iv. p. 135.

member that, after all, he has achieved what the wisest of mankind can seldom outdo, that is, to contribute, in their generation, to the advancement of knowledge. The works of Dr. Shaw, one of the officers of the British Museum, may here be adverted to, as he was unquestionably the writer* of nearly all the zoological descriptions in White's *Voyage to New South Wales*, published in 1789. He has been most aptly termed a " laborious compiler and describer;" † habitually purloining from the works of others, and copying their figures, in popular periodicals of his own; sometimes, although rarely, interspersing them with original articles. He had all the precise technicality, without any of the judgment, of Linnæus. He was, in fact, one of those false disciples of the great Swede, who,—looking to the letter, and not to the spirit, of the *Systema Naturæ*,—brought the reputation of his master into unmerited obloquy; while he imagined he was upholding his fame by a pertinacious rejection of all improvement. His works are scarcely worth enumerating, save as an instance of the mis-direction of good abilities, which occasionally peeped forth, and of the oblivion which will ever attend the writings of those who, for temporary fame, bedeck themselves in the borrowed plumes of others. Such plagiarists, sooner or later, are sure to be detected. We wish that certain compilers of the present day, now in the full tide of

---

* M. Cuvier erroneously attributes the whole to John Hunter, the celebrated anatomist; whereas he merely wrote the account of five of the quadrupeds, and these are neither named nor scientifically characterised.

† Règne Animal, tome iv. p. 156.

their short-lived popularity, would remember this, and desist from similar practices. The writings of Dr. Shaw * may further be cited as a proof of the thraldom in which, at this period, the zoologists of Britain were held by their bigoted devotion to the letter of the *Systema Naturæ.* Nor was it until some years after, when better principles had been established on the Continent, that this unaccountable spell was broken.

(28.) It is worthy of remark, that the very last illustrated publication, of any note, upon Entomology, which appeared in England, and which is arranged in accordance to the Linnæan system, is unquestionably one of the most beautiful and the most valuable that this or any country can boast of. We allude to the two noble volumes upon Georgian Insects †, edited by our late amiable and excellent friend Sir James Smith, the liberal possessor of the Linnæan Museum, and the founder of the Society which bears that name. His labours, indeed, are most conspicuous in botany; but in this work he proves equally conversant both with plants and insects. The plates are the last and best of Harris's performance; and if the reader possesses this work, and

---

* G. Shaw. Vivarium Naturæ; or, The Naturalist's Miscellany, by G. Shaw. London, 1789-90. This came out in 267 numbers, of 3 plates each, nearly all of which are taken from other books, and generally coloured from description. — General Zoology; or, Systematic Natural History. Commenced in 1800, and continued to many volumes. — Zoological Lectures. 2 vols. 8vo.; &c.

† J. E. Smith and Abbott. The Natural History of the rarer Lepidopterous Insects of Georgia, collected from the Drawings and Observations of Mr. John Abbott. London, 1797. 2 vols. folio.

the three volumes of Sepp, his library contains the two best illustrative publications upon Insects that have ever been given to the world. The compilation of Berkenhout*, in 1789, was no doubt useful in its day; and the plates of Lewin †, father and son, are of permanent value, particularly those of the latter. ‡ The volume of Otho Fabricius § must not be omitted; for, independent of the value of its descriptions, it is the only work we possess on the zoology of Greenland. Olivi ||, two years after, wrote in like manner on the marine productions of the Gulf of Venice, with considerable ability, and gave excellent figures of several new *Crustacea* and shells.

(29.) The number of Entomological works, many of them costly and elaborate, which were published on the Continent during the latter part of the last century, were very numerous. Most of them have been noticed in the preceding pages; but several have been omitted in their chronological order; since they would have interrupted the course of our

---

* J. Berkenhout. Synopsis of the Natural History of Great Britain and Ireland. By John Berkenhout, M.D. London, 1795. 2 vols. 8vo.

† W. Lewin. The Insects of Great Britain. By William Lewin, F.L.S. London, 1795. 1 vol. 4to. (containing the Papilios only.)

‡ W. J. Lewin. (1.) Natural History of Lepidopterous Insects of New South Wales. London, 1805. 1 vol. 4to. — (2.) The Natural History of the Birds of New South Wales. London, 1822. Thin folio.

§ Otho Fabricius. Fauna Greenlandica. Leip. 1790.

|| G. Olivi. Zoologia Adriatica, ossia Catalogo ragionato degli Animali del Golfo e delle Lagune de Venezia. Bassano, 1792. 1 vol. 4to.

narrative, while, from being more illustrative than descriptive, they exercised little or no influence in maturing or improving systematic classification. The chief of these we shall, therefore, now enumerate. Some of them exhibit the method of Linnæus; others, that of Fabricius; and a few merely describe the insects that are figured. They may all, however, be considered as belonging to this epoch of the science, when entomologists had no other divisions than orders, genera, and species—when families and sub-families had not been detected—and when, in short, the augmentation of species was considered the most important object of the Naturalist. The beautiful coloured plates of European *Lepidoptera* by Ernst[*], though drawn with little taste, are very faithful, and constitute a valuable set of elucidations of this order: the letterpress is by Father Engramelle, an Augustine monk, and is merely confined to the subjects figured. About the same time, another illustrated work, on the very same subject, was commenced, in German, by Esper[†], a painter of Nuremberg, which continued to be published, at intervals, until it reached five volumes, when it was discontinued; and even these are now

---

[*] Ernst and Engramelle. Papillons d'Europe, peints d'après Nature. Paris, 1779—1793. 8 vols. royal 4to. M. Cuvier is incorrect in stating there are only *six* volumes of this work. See Règ. Anim. tome iv. p. 116. The last part of vol. viii. is excessively rare.

[†] E. J. C. Esper. Die Schmetterlingen in Abbildungen, &c.; or, the Lepidopterous Insects of Europe, figured and described from Nature. Erlang, 1777—1794. 5 vols. 4to. (The *Règ. Animal* erroneously states that there are only four.)

very scarce. Notwithstanding these two voluminous and expensive works on the *Lepidoptera* of Europe, a third still more costly was undertaken by Hübner*, another German draftsman, who seems to have published more on this order of insects than any of his countrymen. A voluminous and costly work was commenced by Herbst and Jablonsky †, in 1782, with the vain attempt of figuring and describing all known insects. It reached to 21 octavo volumes of descriptions, and the plates form the same number of parts; but it was then discontinued. The first two volumes are occupied by the *Coleoptera;* the remainder, by the *Lepidoptera*. The figures of both, however, are chiefly copies; and in the latter division few exotic species will be found, which are not contained in the volumes of Cramer and Stoll. Voet's *Icones* ‡, although somewhat coarsely engraved. are very characteristic of the insects they represent, which are exclusively *Coleoptera;* but the descriptions, and the nomenclature, are worthy only of the age of Mouffet and Petever. The best work upon the

---

* J. Hübner. Der Gamlung Europaischer Schmetterlinge, &c. Augsburg, 1796, &c. 3 vols. 4to. — Beitrage zur Geschichte der Schmetterlinge, &c. Augsburg, 1786—1789. 2 vols. 8vo.

† Herbst and Jablonsky. Natur System Aller, &c.; or, The Natural History of all the known Insects, Indigenous and Exotic. Berlin, 1782—1806. 12 vols. 8vo. 12 do. plates, 4to.

‡ J. E. Voet. Icones Insectorum Coleopterorum Synopsis, Observationibus Commentarioque perpetuo illustravit D. W F. Panzer. Erlangæ, 1794. 4to. — Catalogus Systematicus Coleopterorum, Figuris coloratis. 2 vols. in cases. (Haworth's Lib.)

European *Hemiptera* is unquestionably that of Wolff*, published in *fasciculi* or parts ; but we know not, with certainty, how many have appeared. Wolff adopts the Fabrician system ; and both in his descriptions and figures he is very superior to the generality of Iconographers. In the same year, Schellenberg †, a painter of Zurich, figured many insects of the same order inhabiting Switzerland ; and subsequently published an indifferent work upon the two-winged genera, or *Diptera*. The Entomological plates of our countryman Donovan, although frequently too highly coloured, and not sufficiently accurate in the more important details, are often elegant, and frequently useful, especially those contained in his three quarto volumes, where a great number of species are delineated for the first time. Little, however, can be said in praise of his works on other departments of British Zoology, the colouring of which is gaudy, the drawings generally unnatural, and the descriptions unsatisfactory.‡ The works of Uddman, Barbut, Bradley, Martyn, Marsham, and a few others, published at different periods, are too subordinate to deserve a particular notice.

---

* I. F. Wolff. Icones Cimicum, Descriptionibus illustratæ. Erlangæ, 1800. 4to.

† J. R. Schellenberg. Cimicum in Helvetiæ Aquis et Terris degens Genus. Turici, 1800. 8vo. — Genres des Mouches Diptères. Zurich, 1803. 8vo.

‡ Ed. Donovan. (1.) The Natural History of British Insects, explaining them in their several States, illustrated with coloured Figures. London, 1792—1820. 16 vols. royal 8vo.— (2.) General Illustration of Entomology; being Epitomes of the Insects of China, India, and New Holland. London, 1798—1805. 3 vols. 4to.

There are some authors, however, not yet mentioned, whose names occupy a superior station in the Entomological history of this period. The first of these is the laborious Panzer*, who began publishing, in 1796, a collection of figures and descriptions of the Insects of Germany. There does not exist, among all those we have enumerated, a more accurate or a more useful work. The figures are drawn and etched by the famous Sturm, the best entomological artist on the Continent; and are simply, but accurately, coloured; while the descriptions, although frequently too short, are written by the hand of a master. The system of Fabricius is followed; and the work altogether is highly essential to every one who writes upon the entomology of Europe. A valuable pamphlet, by professor Petagni †, on the Insects of Lower Calabria, a fruitful field for the entomologist and which has hitherto been very little explored,— appeared in 1787; and the *Institutions* of the same author contributed very much to spread a taste for this science in Italy, whose entomology had already been ably illustrated by Professor Rossi ‡ of Pisa.

---

* G. W. F. Panzer. Faunæ Insectorum Germanicæ Initia; or, Deutschlands Insecten. In 109 fasciculi, each containing 24 plates and their descriptions. In 12mo. Nuremb. — Index Entomologicus, pars prima, Eleutherata. Nuremb. 1813. 1 vol. 12mo.

† V. Petagni. Specimen Insectorum ulterioris Calabriæ. Neapoli, 1786. 4to. — Institutiones Entomologicæ. Neapoli, 1792. 2 vols. 8vo.

‡ P. Rossi. Fauna Etrusca; sistens Insecta quæ in Provinciis Florentinâ et Pisanâ præsertim collegit Petrus Rossius. Liburni, 1790. 2 vols. 4to. — Mantissa Insectorum Etruriæ Pisis, 1792. 1 vol. 4to.

No additions had been made to the natural history of Sweden, since the publication of the delightful *Fauna* of that country by Linnæus; but, in 1798, G. Paykull*, one of the ministers of the king, commenced an elaborate work on the coleopterous insects of his native country, which extended to three volumes. The last name we shall mention of this entomological era, is that of Laspeyres, who has most beautifully and ably illustrated and described the European *Sesiæ*, in a work which must long remain a model for future monographers.† Finally, we may here mention Mr. Dillwyn's ‡ Conchological work, which, like that of Villers in regard to insects, is an attempt to notice all the species of recent shells to the Linnæan genera.

(30.) We have now brought the series of Linnæan writers to a close, with the exception of one, whose laborious and voluminous work seemed necessary to convince the strict adherents to this school, of the absolute necessity of a reform in systematic arrangement; and that it was utterly impossible any longer to delay those improvements which Linnæus, by his own example, had so forcibly inculcated. These truths were forced upon the conviction of the most preju-

---

* G. Paykull. Fauna Suecica, Insecta. Upsaliæ, 1798. 3 vols. 8vo. This date (for there is none on the titlepage) is prefixed to the end of the preface. M. Cuvier erroneously gives 1800 as the year of its first publication.

† J. H. Laspeyres. Sesiæ Europææ. Iconibus et Descriptionibus illustratæ. Berolini, 1801. 4to.

‡ F. W. Dillwyn. A Descriptive Catalogue of recent Shells, arranged according to the Linnæan System. London, 1817. 2 vols. 8vo.

diced by the appearance of the thirteenth edition of the *Systema Naturæ*, " enlarged and reformed" by Dr. Gmelin.* It is unnecessary to say that the worthy editor was one of those admirers of the great Swede, who departed not from the letter of his master, and that consequently he admits scarcely any of the improvements made by his more judicious followers; nor has Gmelin (probably from not having had the use of a rich national museum, like that of France) endeavoured to unravel the innumerable errors of his predecessors. His sole object was to concentrate their labours; and in this he has shown, if not judgment, at least great and singular research and industry. It unfortunately happens, that the odium of those errors, which he had probably no means of detecting, has been thrown upon him, and his real merits completely overlooked. It may fairly be questioned whether, in regard to the nomenclature of species, there are not to the full as many errors in the *Règne Animal*, as there is in the compilation before us. The time, in fact, had even then arrived, when it became utterly impossible for any one individual, who undertook to illustrate the whole animal kingdom, to examine the characters and the synonyms of species: he *must*, in numberless instances, repose on the opinions of others; and consequently must lay himself open to the charge, however severe, of perpetuating error. Gmelin, as a compiler, and he pretends to nothing beyond, is neither inferior to those who preceded or to those who followed him;

---

* J. F. Gmelin. Caroli a Linné, Systema Naturæ per Regna Tria Naturæ. Editio decima tertia, aucta, reformata. Curâ J. Fred. Gmelin. Lipsiæ, 1788. 3 vols. (in 10 parts) 8vo.

and he has this merit, which no others possess, that he gave a much fuller compilation upon all that was then known of the animal kingdom, than is to be met with in the records of our science.

(31.) Having now traced the progress of that school, which, under the guidance of Linnæus, commenced about the year 1754, we must carry the attention of the reader back to the middle of the last century, when there arose, as we have before stated, a formidable rival to the luminary of the North, in the celebrated Buffon, who, with a pertinacity unworthy of his talents, set out with despising all system, and all technical helps to the communication of knowledge; and thus formed a school of his own. It is the character and the progress of this school which we are now to trace. That it had plausible, and even valid, grounds for dissent, is readily admitted; but had there been a cordiality of spirit between the respective founders and their disciples, their talents might have been united without prejudice to either, and science would have advanced, probably, in a double ratio to that in which it really proceeded. It is easy to despise that which requires trouble to learn ; and to call an animal by a name of our own, regardless of that by which it is known to the world, is obviously neither a proof of sound sense or of good judgment. Yet such was one of the characteristics of the school of Buffon, who set out with rejecting the classic names of all his predecessors, substituting for them a barbarous nomenclature, composed of words half savage, half French, without meaning or without sense. Natural history, under such a principle, would have become unintelligible; and what should have been the language of science, would, had the plan succeeded, have been turned into an unintelligible jargon, the words of which, if they could be so called, in many instances would have been almost unutterable. It was the object of Buffon to write an historical biography of every animal — while that of Linnæus was to express its peculiar characters in as few words as possible. It is quite clear that both these objects could be combined, for the one interferes not with the other; but the pride of Buffon would not permit him to show, by his writings, that he approved of any thing which came from Linnæus; and his disciples, of course, followed his example. On the other hand, it must be admitted, that the dry and technical style of the *Systema Naturæ* (the inevitable consequence of the condensing system Linnæus went upon) was exceedingly distasteful to all but professed naturalists. There are a thousand circumstances of popular interest in the economy of animals, which yet are not necessary to be touched upon in a bare descriptive catalogue of distinctions. It is the happy art of throwing these circumstances into a connected history, which gains popular applause; and although such narratives are not always the most valuable, they are unquestionably the most generally interesting. Nor are they devoid of interest even to the philosophic zoologist: on the contrary, the habits and instincts of an animal are as essential to determine its true relations to others, as are its external or internal structure: for as, in the moral world, we judge the character of a man, not from a single act, but by the tenor of his life, so, in the natural world, it is necessary to the right understanding of the station which an animal holds in the scale of creation, that all its characteristics are known, either from actual observation, or by analogical arguments drawn from its general structure. With these preliminary remarks, let us now take a rapid survey of the writers who belong more or less to this descriptive school, nearly all of whom are countrymen of their master.

(32.) The rapidity with which succeeding editions of the works of Buffon were called for, almost equalled the avidity which was manifested to possess the *Systema Naturæ;* and both had a most extensive circulation. It is unnecessary for us, however, to enumerate the various reprints of these works, some one of which are in the hands of almost every naturalist. Buffon's work has been more than once translated into English, but hitherto by no one at all qualified for the undertaking: the translation by Woods is, probably, the best. His History of Birds was illustrated by a separate publication, at the cost of the government, but without letterpress, generally termed the *Planches Enluminées*. These consist of one thousand and eight coloured plates of birds, printed both in folio and in quarto. The execution of these plates has been much over-rated; although they were doubtless the work of the best artists then to be met with in France: they are very inferior to those of Edwards; and the best that can be said of them is, that they are recognisable. That they even still continue to be essential for purposes of reference, is entirely owing to the enormous expense of publishing such a voluminous collection of plates.

We have, indeed, commenced the publication of a similar collection*, now so much wanted for our public libraries and institutions; but it is highly probable that so few copies will be printed, that the entire work will only be in the possession of the original subscribers; and thus the main object of the undertaking will be but very partially accomplished. Bonnet †, the celebrated philosopher of Geneva, influenced, probably, by the example of Reaumer, published two volumes upon insects. But both these authors were surpassed by the illustrious Baron de Geer ‡, who, in the year 1752, gave to the world his first volume of *Mémoires,* which he subsequently extended to six others. Every entomologist who has had occasion to mention this invaluable work, concurs in bestowing upon it their unqualified praise; not only for the admirable and interesting details it contains on the structure and habits of the insects described, and the beauty of the investigations it narrates, but for the just and comprehensive views it unfolds on natural arrangement. It is a subject of much regret that this work is of such exceeding rarity as to be quite unprocurable. We have never been successful in meeting with a copy for sale; and although it was soon

---

* W. Swainson. Ornithological Drawings, in Geographic Series. Series I. The Birds of Brazil. Parts 1, 2, and 3. London, 1834. Royal 8vo., published quarterly, 12 plates in each.

† C. Bonnet. Traité d'Insectologie. Paris, 1745. 2 vols. 8vo.

‡ De Geer. Mémoires pour servir à l'Histoire des Insectes. Stockholm, 1752—1778. 7 vols. 4to.

translated into German, it has never been put into an English dress.

(33.) M. Reaumer, whose family and connections were high, besides being attached to entomology, possessed a very noble collection of birds, and this was no doubt the chief inducement to M. Brisson, the curator of his museum, for commencing his *Ornithologie* *, wherein he comprehends all the well-authenticated species then known, whether in his patron's museum, or described in books. The chief, and indeed the only, merit of this voluminous work is the extreme exactitude of the descriptions; for the figures are scarcely superior in drawing to those of the *Planches Enluminées*, and, being un-coloured, are less recognisable. It is curious to observe a trait of *littleness* in the mind of this otherwise estimable writer, which clearly shows the feelings of jealousy, if not of hostility, with which the writings of Linnæus were then viewed in France. M. Brisson departs so far from the school of Buffon, as to arrange birds in a system of his own; and he even goes so far as to give them names, and specific characters, in Latin: but although he quotes the writings of Linnæus, he will not even mention his specific names, and scarcely adopts any one of his genera. With all these defects, the volumes of Brisson are nevertheless still valuable, as containing minute descriptions of birds then considered new,

---

* M. J. Brisson. (1.) Le Règne Animal, divisé en 9 Classes. Paris, 1756. 1 vol. 4to. (2.) Ornithologie; ou, Méthode contenant la Division des Oiseaux, en Ordres, &c. Paris, 1760. 6 vols. 4to. (3.) Ornithologia, sive Synopsis Methodica sistens Avium, &c. Lugd. Bat. 1762. 2 vols. 8vo.

and as supplying those details which were not consistent with the conciseness of the Linnæan plan.

(34.) The name of Adanson * is recorded both in botany and zoology; not so much for the value of his works, as from his being among the first of those, who, like our countryman Lister, endeavoured to arrange shells with some regard to the structure of their animals. His love for natural history carried him to the coast of Senegal, the shells of which he has described, and tolerably well figured, in a separate volume, still of great value to the conchologist. We may here observe, that Adanson, like his master Buffon, was a declared enemy to the regularity and system which governed the Linnæan nomenclature; and that our author (proceeding on the plan of this school) calls his shells by Negro-French names. Thus, on the *Voluta Cymbum* of Linnæus, he bestows the name of *Yet;* the *Voluta Cymbiola* is to be *Phelan;* and the *Marginella lineata* is called a *Bobi!* There is, in short, no end of such names as Lupon, Bitou, Salar, Mafau, (we take them at random,) Minjac, Sakem, Sadot, Pakel, and innumerable others. This is the jargon which Buffon, influenced by his regard for elegance of diction and of phraseology, strove to substitute for the classic and expressive nomenclature of Linnæus! One is really surprised, in these days, to contemplate such folly, as proceeding from reasonable beings; did not prejudices equally great, but often far more hurtful, meet us at every step in our journey through life.

---

* M. Adanson. Histoire Naturelle des Coquillages du Sénégal. Paris, 1757. 1 vol. 4to.

We have already mentioned the entomological works of Schæffer, which might with equal propriety be classed in the present enumeration; for though their author admitted genera, he rejected specific names, and described his insects in the obsolete style of the early entomologists. The volume of Duhamel, upon Ichthyology, is now chiefly valuable for its figures: while those of Sonnerat*, wherein a large number of Indian birds are tolerably described, but wretchedly figured, are of little use; there are no determinate or scientific names; and the descriptions puzzle, rather than assist, the ornithologist. Sonnini †, one of the engineer officers of the French army in Egypt, is chiefly known, as a naturalist, by the edition of Buffon which bears his name. It is the most copious we have seen, and the best; being enriched with many original observations of the author made upon the spot upon the birds of Cayenne. In his Egyptian narrative, which was translated into English, will be found descriptions and figures of many of the new fish of the Nile.

(35.) But the greatest ornithologist of this school is the celebrated Le Vaillant, an enthusiastic tra-

---

* Sonnerat. (1.) Voyage à la Nouvelle Guinée. Paris, 1776. (2.) Voyage aux Indes Orientales et à la Chine, depuis 1774 jusqu'en 1781. Paris, 1782. 2 vols. 4to.

† C. S. Sonnini. (1.) Histoire Naturelle des Oiseaux, par Le Clerc de Buffon. Ouvrage formant une Ornithologie complète, par C. S. Sonnini. The editor has been at considerable labour in adding all the synonyms and Latin names. Paris, An. XII. (1798, &c.) 28 vols. 8vo. (2.) Voyage dans la Haute et Basse Egypte. Paris, 1799. 3 vols. 8vo., and atlas of plates.

veller, and a most accurate observer of nature. He expended his entire fortune in producing the most magnificent series of works, upon his favourite study, we possess; and he has been the chief guide to MM. Temminck, Cuvier, and nearly all the moderns, in respect to the genera and families of African birds. Le Vaillant*, unfortunately for his own fame, was a rigid disciple of Buffon. He affected to despise system, and would only use French names. The consequence has been that those who came after him have had all the honour of incorporating and classifying his discoveries in the regular systems; and groups which were first distinguished and pointed out by himself, are now only known by the names given to them in the *Règne Animal*, and other works. The plates of his three last works are exquisite, having been made from the drawings of Barraband, the best ornithological painter France has ever produced. We may here introduce the name of the Spanish naturalist, Don Felix de Azara, who investigated, with great ability and unwearied zeal, the quadrupeds and birds of Paraguay, of which province he was the governor. Azara, like Le Vaillant, rejected system; but although his descriptions are not only

---

* F. Le Vaillant. (1.) Histoire Naturelle des Oiseaux d'Afrique. Paris, 1799. 6 vols. folio or quarto. (2.) Histoire Naturelle des Perroquets. Paris, 1801. 2 vols. folio or quarto. (3.) Histoire Naturelle des Oiseaux de Paradis, des Rolliers, des Toucans, et des Barbus. Paris, 1806. 1 vol. imp. folio. (4.) Histoire Naturelle des Promerops et des Grêpiers. Paris, 1807. folio. (5.) Histoire Naturelle d'une Partie d'Oiseaux de l'Amerique et des Indes. Paris, 1801. 4to.

accurate, but masterly, yet, from not being referred to any of the modern genera, or accompanied by plates, they are, in numberless instances, perfectly useless, from the impossibility of determining the systematic characters of the animal described. This is greatly to be lamented, for he is the only writer on the zoology of South America who has recorded the economy and habits of the animals he describes. The entomological memoirs, collected into the volume of Fuessly *, are partly in the style of narrative adopted by Reaumer, and partly systematic; but both are interesting and instructive, and the figures well executed.

(36.) The narrative style of treating natural history, adopted by Buffon and his immediate followers, however interesting and popular, was soon found to be quite inconsistent with the study of nature as a science; and even the most eminent of his own countrymen, when the fever of admiration had somewhat subsided, began to see the impossibility of going on without a more orderly method of arranging their discoveries, and of communicating their knowledge: some, therefore, adopted the Linnæan or the Fabrician system, or invented one of their own; while others, of a higher order, perceived that not only *system* was to be implicitly followed, but that a much more complicated one than that of Linnæus was necessary. Hence arose a new school of zoologists in France; who not only embraced the spirit of the Linnæan mode of arrangement, but

---

* J. G. Fuessly. Archives de l'Hist. des Insectes. Winterthhour, 1794. 4to.

proceeded to other and much more numerous combinations and divisions. This school was founded by three eminent men, all of whom have disappeared from the ranks of science within the last two years. M. Lamarck* undertook the investigation of the invertebrated animals; M. Cuvier†, the vertebrated; and M. Latreille‡, the class of *Annulosa* or of Insects. The systems respectively invented by these able zoologists will be examined in some detail during the course of this publication, and it will therefore be unnecessary, in this place, to investigate their merits. We have also come to that era of the science, whereof many of the chief actors are now living; and whose works cannot, with propriety, be spoken of with that freedom (and, we hope, with that impartiality) we have hitherto done, and which is expected from the historian of times that are past. We shall, therefore, merely state the chief characteristics of that school which succeeded, in France, to that of Buffon, and briefly enumerate the leading works which it has produced.

(37.) It may naturally be supposed, that since the time of Linnæus, our knowledge of nature had been vastly extended; so that the species had been more than quadrupled. Hence arose the necessity of instituting a proportionate number of new genera,

* Lamarck. Hist. Nat. des Animaux sans Vertèbres, par M. le Chevalier de Lamarck. Paris, 1815—1823. 7 vols. 8vo.

† Cuvier. Le Règne Animal, distribué d'après son Organisation. Paris, 1817. 4 vols. 8vo.

‡ P. A. Latrielle. Genera Crustaceorum et Insectorum. Paris, 1806. 3 vols. 8vo.

in all the branches of zoology; and of introducing several intermediate groups between those which Linnæus termed orders and genera. But all this could not be done without a more rigorous investigation, into the structure of animals, than was formerly required. The study of internal comparative anatomy was therefore called in to aid zoological classification; without which it was found impossible to understand aright the true nature of many of the molluscous animals of Linnæus, or of those stupendous remains of extinct animals found in a fossil state. Now, had the employment of this new science been limited to such and similar cases which really required its aid, from the insufficiency of external distinctions, all would have been well; but the leaders of this school, more especially M. Cuvier, delighted with the success that attended their first researches, proceeded at once to proclaim that internal anatomy was the only sure basis of the natural classification; quite forgetting the fact, that external structure was just as important for this purpose as internal anatomy; and that the one, in most cases, is but an index to the other. To bring this home to the conviction of every one, and as an illustration of our meaning, let us look to a gallinaceous bird — the common fowl. If, by its general shape and external organisation, we can judge of its habits, its mode of life, the nature of its food, and of its powers of locomotion; and if the knowledge so gained, is quite sufficient for every purpose of recognition and of classification; where would be the necessity of proceeding further? why should these definitions be burthened with others, taken from the internal anatomy? why call in the aid of another science to make that object more perfectly known, which was before sufficiently plain for all the purposes of recognition? If, as it has been asserted, natural arrangement depends upon internal anatomy, how do we know that it is not equally dependent upon chemistry? Has this theory led to the discovery of the natural system? or to any one of those laws by which such a system may be supposed to be regulated?— Certainly not. The law, as it has been termed, of the *condition of existence**, is no more than that every animal is constructed according to the functions it is destined by nature to perform. Now, so far from this, as some have insinuated, being a modern discovery, it was well known to Aristotle; and is a truth apparent to the most superficial observer. It must be admitted, however, that M. Cuvier is the only one of this school who has attached to this theoretic principle of internal organisation so much undue importance: an error he was obviously led into from the splendid success which attended its use in his researches on the fossil bones; where, indeed, a complete knowledge of comparative anatomy was absolutely indispensable. It is not maintained that a knowledge of internal anatomy is superfluous to the zoologist; but that it is quite redundant (and therefore unnecessary), where all that is essential to be known of an animal can be learned from external organisation. With the exception, therefore, of M. Cuvier, the systems of his celebrated cotemporaries may be said to make

---

* Règne Animal.

a much nearer approach to that of nature than any which had preceded them, because their groups are founded upon *general* considerations. The orders of Linnæus were subdivided and remodelled, the nature of the molluscous, radiated, and annulose animals defined with great skill, and every part of the animal kingdom was minutely analysed and more correctly defined. The result of all this was collected into the *Règne Animal* of the celebrated Cuvier, which may be termed the *Systema Naturæ* of this era, and which certainly contains a greater mass of zoological information than is to be found in any modern work. It has, with justice, been compared to a mine of information, rich both in the discoveries of the author and of his cotemporaries. " But the disposition and ability to make use of this one, to give it the proper form and polish, is not, it seems, a necessary concomitant to skill in extracting it, or to the patience required before it could have been collected for use. At least, it is but too visible, and has been too often and too justly remarked, that no person of such transcendent talents and ingenuity ever made so little use of his observations towards a natural arrangement as M. Cuvier."* The amateurs of zoology, in this country, ever prone to judge in extremes, after overlooking the labours of this great man for nearly

---

* Horæ Entomologicæ, p. 326. And yet it has been stated, in a public *éloge* pronounced in this country, that Cuvier eminently possessed these powers of " legitimate and inductive generalisation" in *arranging the animal series*, in which, as Mr. M'Leay truly observes, he was so notoriously deficient.

seventeen years, in their bigoted devotion to the letter of the *Systema Naturæ*, have now flown to the opposite extreme. They have invested his memory with a universality of talents almost superhuman ; and are now ready to bow to his authority with that blind and implicit homage they formerly paid at the shrine of Linnæus. It may, therefore, surprise such persons, to be told that, in the investigation and knowledge of recent quadrupeds, M. Cuvier has been fully equalled by the illustrious Geoffroy St. Hilaire ; that his system of Ornithology is inferior to that of Temminck, and is withal so defective, that it has called forth an exposition from one of the first zoologists of Europe*; in short, that it has never made one convert. That in the anatomy of the *Mollusca* and soft animals, he was not only preceded, but greatly surpassed, both by the celebrated Poli and the incomparable Savigny † ; while, in their arrangement, he is confessedly inferior to Lamarck ‡; and finally, that the whole of the entomology of the *Règne Animal* is avowedly from the pen of Latreille. If the fame of M. Cuvier, therefore, reposed upon his talents as a zoologist, or as a classifier, that fame would not outlive the present day, for his system has been already shaken to its very foundations. No ; it is the transcendent genius he has shown as a geologist and comparative anatomist, in his splendid

---

* Sulla Seconda Edizione del Regno Animale del Barone Cuvier. Osservationi de Carlo Luciano Bonaparte, Principe de Musignano. Bologna, 1830. 8vo.

† J. C. Savigny. Mémoires sur les Animaux sans Vertèbres. Paris, 1815, 1816. 2 parts, 8vo.

‡ See also, Horæ Entom. p. 837. &c.

theories, and his fossil investigations, that will perpetuate his name so long as those sciences are cultivated: and they will be mentioned with admiration, when the *Règne Animal*, for all purposes of philosophic or natural arrangement, will serve only, like the *Systema Naturæ*, to mark the period of a bygone era. It is with deep regret that the Christian philosopher traces another peculiarity in this school; which applies, more or less, to the greatest number of the works it has produced. In perusing the discoveries they contain, brilliant and elaborate though they be, we look in vain for that pure spirit of religious belief which breathes in the writings of the gentle Ray, or those bursts of lofty praise and enthusiastic admiration of Nature's God which break forth from the great Linnæus, and which irradiate all that he ever wrote. A cold, ill-concealed spirit of materialism, or an open and daring avowal of wild theories, not more impious than they are absurd, attest, too unequivocally, the infidelity that attaches to some of the greatest names in modern zoology which France, or indeed any other country, has produced.

(38.) The era now before us, although of short duration, includes a host of learned, accurate, and accomplished zoologists; most of whom are happily still living, and still investigating. England may claim the merit of first originating this analectic mode of investigating nature; for the celebrated work of our pious and venerable countryman, Mr. Kirby*, was

---

* W. Kirby. Monographia Apum Angliæ. Ipswich, 1802. 2 vols. 8vo.

undoubtedly the germ of that revolution in entomology subsequently effected by Latreille*, whose labours in this department are immense. Geoffroy St. Hilaire, and the two Cuviers†, prosecuted the study of Quadrupeds in France, while Illiger‡ was doing the same in Prussia. The exquisite and elaborate works of Poli§ on the comparative anatomy of the *Mollusca*, is alone sufficient to immortalise a name; and this unrivalled publication led the way for the valuable memoirs on the same class by Cuvier, which were subsequently collected into a volume. ‖ Lamarck, well characterised as the most accomplished zoologist of this era¶, took up the whole of the invertebrated animals: while a series of splendid illustrations in folio, by Temminck, Desmarest, Vieillot, Audebert**, and many others of a still later date,

---

* P. A. Latreille. Genera Crustaceorum et Insectorum. Paris, 1806, 1807.

† MM. Lacépède, Cuvier, and Geoffroy St. Hilaire. Ménagerie du Muséum d'Histoire Naturelle. Paris, 1804. 2 vols. 4to.

F. Cuvier and Geoff. St. Hilaire. Histoire Naturelle des Mammifères. Paris, 1819—1822. folio, in numbers.

‡ C. Illiger. Prodromus Systematis Mammalium et Avium. Berolini, 1811.

§ J. X. Poli. Testacea utriusque Siciliæ. Parmæ, 1795. 2 vols. imp. folio.

‖ Cuvier. Mémoires pour servir à l'Histoire et à l'Anatomie des Mollusques. Paris, 1817. 1 vol. 4to.

¶ Horæ Entom. p. 328.

** C. I. Temminck. Histoire Naturelle Générale des Pigeons. Paris, 1808. folio.

A. Desmarest. Histoire Naturelle des Tangaras, des Manakins, et des Todiers. Paris, 1805. folio.

L. P. Vieillot. Histoire Naturelle des plus beaux Chan-

attest the progress which had now been made in zoological painting. The birds of Europe were most ably investigated by M. Temminck*, who has also written largely upon the Gallinaceous order. Voluminous dictionaries of natural history, in all its branches, followed each other in rapid succession; until at length the *Règne Animal* became as insufficient a vehicle for concentrating this vast accession of knowledge, as was the *Systema Naturæ* at the death of Linnæus.

(39.) While the details of zoology were thus prosecuted in France with an ardour and a success perfectly unexampled, a feeling arose in the minds of a few eminent men of other countries, that the time had now arrived when an effort might be made to generalise the innumerable facts thus elicited; and to reconcile, in some measure, the conflicting systems that were following " thick and fast" upon each other. The science of zoology, up to this period, had assumed no appearance of collective symmetry. Every department had its own independent system; and although great order and regularity had been introduced into each, yet all the divisions

---

teurs de la Zone Torride. Paris, 1805. folio. — Histoire Naturelle des Oiseaux de l'Amérique Septentrionale. Paris, 1807. 2 vols. folio.

J. B. Audebert. Histoire Naturelle des Singes et des Makis. Paris, 1800. folio.

J. B. Audebert et Vieillot. Histoire Naturelle des Oiseaux Dorés; ou, à Reflets Métalliques. Paris, 1802. 2 vols. folio and 4to.

* C. I. Temminck. Manuel d'Ornithologie; ou, Tableau Systématique des Oiseaux qui se trouvent en Europe. Paris, 1820. 2 vols. 8vo.

of families, genera, &c. merely reposed upon the arbitrary opinion of their founders. Nay, so destitute was zoology of any fundamental law, applicable alike to all its various departments, that the question was not yet settled, as to the rule of natural progression; was it linear? was it compound? or was it so interwoven, like the meshes of a net, as to defy all unravelment? The idea of a simple scale in nature had long been discussed, and finally abandoned. But while these lofty speculations engaged not the attention of M. Cuvier, his fellow-labourer, M. Lamarck, must have long pondered upon them, for he it was who first intimated the existence of a *double* series, which, setting out in opposite directions from a given point, met together at another. Nearly at the same time, Professor Fischer, a celebrated zoologist of Russia, unacquainted, apparently, with the opinion of Lamarck, perceived the tendency of these two series to form a circle of their own, and announced the fact in 1808. But these obscure intimations, unsupported by demonstration, can only be said to have been verified by analysis when the first part of the celebrated *Horæ Entomologicæ* was given to the world, in 1819,—a work which, for its originality and profound research, has never yet, in this science at least, been equalled. Whether its accomplished author derived the first idea of that circular progression of affinities which he establishes, from the idea of Lamarck, is unknown, and hardly worth enquiring into; but it seems certain that he was unacquainted with the opinion of M. Fischer, just alluded to.*

---

* Linn. Trans. vol. xvi. p. 10.

Four years after, a celebrated botanist of Germany, E. Fries, equally ignorant of the previous discovery of M'Leay, announced the same fact as manifested in the vegetable world, and which he demonstrates by a much more extensive analysis than had been given, in regard to insects, in the *Horæ Entomologicæ*. It is not the least remarkable circumstance connected with this splendid discovery, that four individuals, in different countries, and unknown to each other, should all have directed their studies to the same object, and that all should have arrived at the same result: thus establishing, what had never yet been done, the existence of at least one universal law in natural arrangement, and thus raising zoology, for the first time, to the rank of a demonstrative science.

(40.) This era, then, of our science has just commenced, and here must we close our sketch. It is not expedient that the historian should continue his narrative, when he himself becomes an actor upon the stage. We therefore resign to another pen the task of recording the passing events in the history of our science, and proceed to trace its influence on the moral and practical duties of life.

PART II.

ON THE GENERAL NATURE AND ADVANTAGES OF THE STUDY OF NATURAL HISTORY

CHAPTER I.

INTRODUCTORY REMARKS. — WHAT NATURAL HISTORY IS, — IN A GENERAL SENSE, AND AS NOW RESTRICTED. — DIVISION OF THE SUBJECT. — REFLECTIONS ON NATURE AND ART. — DISTINCTIONS, AND OBJECT OF THE STUDY.

(41.) COULD we suppose man had never known evil, — that he had continued, as at first created, a terrestrial, yet an immaculate being, alike a stranger to the bad passions and the inordinate desires that now agitate him, — what pursuits, may we suppose, would occupy his time? or upon what subjects would he exercise those powers of reason by which he is united to the spiritual world? The answer is obvious. The works of God, as manifested in all visible nature, would be his only study. Surrounded by innumerable objects attractive by their beauty, wonderful by their construction, or interesting by their economy, his days would be spent in surveying the material world; — his heart enlarged, and his reason exercised, in meditating on all that he saw. Every new discovery would increase his veneration for the Divine Author of such wonders; and although placed upon earth, his contemplations would be those of the inhabitants of heaven. Such is the reply suggested by reason, to our previous question; and such, does inspiration assure us, was the occupation of the parent of mankind. " And Adam gave names to all cattle, and to the fowls of the air, and to every beast of the field." It is fit that the study of nature should be coëval with the creation of man. Though his spirit has been changed, — though care and trouble, those thorns and thistles of his present state, entangle and distract him, and he is called to the discharge of moral and social duties, — yet this remnant of primeval happiness is still left to him. The volume of nature, with all its variety and beauty, still lies open for his perusal; and in those short hours snatched from the stirring excitations of the troubled world, he may still turn aside, and consider the lilies of the field; and he may read, in the metamorphoses of the butterfly, the change that awaits himself.

(42.) All knowledge may evidently be referred to one or other of the following divisions: — First, such as regards the works of God; and secondly, such as emanates from the inventions of man. As the former is the most noble and the most intellectual, so is it the most comprehensive; since it regards not only the natural objects which surround us, but the internal composition of those objects, and the laws by which the phenomena of nature are regulated. Natural History, therefore, in this its most extended sense, may be considered as embracing the study of matter, whether ponderable or imponderable, whether the objects we contemplate are visible to the eye, palpable to the touch, or invisible agents known only by their effects.

(43.) But the human mind, limited in its powers, is compelled to relinquish the study of universal nature, and to confine its researches to distinct portions. Hence has originated the necessity of instituting those numerous divisions in natural philosophy, respectively assigned to the astronomer, the chemist, and the physiologist. These pursuits, like others of a subordinate nature, are no longer considered as forming a part of natural history, properly so called; although, in a general sense, they strictly and exclusively emanate from the study of nature. Geology, in like manner, separates itself as a distinct department; not because it merely embraces terrene objects, but because it relates more to the situation, than to the analysis, of the component parts of our globe. Its chief business is to trace and explain the changes and revolutions which have happened to the earth; but not, like mineralogy, to determine the primary elements of which it is composed. Natural history, thus restricted, may, in a philosophic sense, be termed the study of ponderable matter, or, to state this definition in more popular language, it is the province of natural history to embrace all that concerns the three great divisions or kingdoms of nature,—the animal, the vegetable, and the mineral. Such is the view which, in common with some of the highest authorities, we propose to take of this science. And although our subsequent remarks will chiefly relate to the animal kingdom, they may be considered, in most cases, equally applicable both to the vegetable and the mineral.

(44.) Let us now consider the objects, whether immediate or remote, which this science comprehends, and the advantages that may be expected to result from its study. We shall regard it, *first*, as intellectual; *secondly*, as recreative; and *thirdly*, as affecting the ordinary business of life.

(45.) It may be received as an indisputable truth, that no studies are so well suited to the intellectual powers of man, as those that relate to the forms and the phenomena of Nature. Between these, and such as are confined to human skill or to human erudition, there is this remarkable difference: that in the former we contemplate things which, in themselves, are perfect, because they emanate from the Fountain of Perfection; whereas in the latter our attention is absorbed in things which, at the best, are imperfect, however exquisite may be the art which produced them, or however learned or acute may be the labours of their authors. The painter or the sculptor may delight us by the faithfulness of their delineations; the poet may please us by the harmony of his verses; the historian instruct us by the narrative of circumstances and persons before unknown to us. But all these subjects, however interesting or pleasing, are alloyed with that imperfection and unsatisfactoriness which enter into every human performance: they chiefly, if not exclusively, refer our thoughts to their authors; or if we even discover no imperfection which mars the painting or the statue — no word which destroys the harmony of the poet's verse, or no imagery which deforms it, if our feelings are neither pained nor indignantly roused by the narrative of the historian, still we rise from the subject with the melancholy conviction that these things are perishable; that the cunning hand of the artificer, and the master-spirit of the narrator, has either passed away or will soon be laid in the dust; and that these records of their skill or of their genius may be lost or destroyed by one of those thousand accidents which have already swept into oblivion so many similar productions There are few contemplative men, after viewing those celebrated fragments called the Elgin Marbles, who have turned from them without some such feelings as those we have described. Our wonder, indeed, is excited at the exquisite skill which is still so conspicuous in these relics; but the sight of decay and dilapidation is at all times melancholy. We are not only reminded of the instability of every thing human; but a vague suspicion must cross the mind even of the most successful, that his own labours, upon which he fondly builds his hopes of deathless fame, may share the same fate; and that a time may come when not only his works, but his very name, may be blotted from the records of future generations.

(46.) If, on the other hand, we turn to those studies which more immediately concern Nature, we find a marked difference both in the facts and in the deductions. Here we have to do with things immutable, and with objects perfect in structure. Our mental perceptions are employed in contemplating phenomena which have remained, for the most part, unchanged from the beginning, and will continue unchangeable so long as the laws which govern the universe remain in force. Here is no extinction of the species, no power of detecting imperfections, no regrets at the insufficiency of the artificer, no lamentations that such things will pass from the earth, and be forgotten. Nature is ever the same — ever young — ever the handmaid of One who cannot err. Her operations in the physical world were the same a thousand years ago as they are now; and if the works of her commentators are no more remembered, this oblivion originates not in any change in the things they treated of, but in the errors or insufficiency of the describers.

(47.) The mutability proverbially belonging to human learning, has been indiscriminately applied both to arts and sciences; whereas it is by no means equally shared between both, nor is it so universal as some would lead us to imagine. Art more correctly implies physical dexterity: science, on the contrary, is purely intellectual. The first cannot exist in any eminent degree, without the second; but science requires not the auxiliary help of her sister. The one is transient, and, however great, dies with its possessor. The painter cannot bequeath to his disciple that skill which it has cost him his life to attain; the poet cannot infuse his "unutterable thoughts" into another before his death; nor can the musician, while he transfers his instrument, delegate also the pathos or the dexterity which gave it utterance. The degree of perfection to which each of these artists has attained, dies with its possessor; and those who succeed him have to begin, themselves, at the foot of the ladder, and not from that height which their predecessors had reached. Hence it is, and the inference is remarkable, that, in those pursuits which more immediately regard art, mankind has but little, if at all, advanced, during many centuries. Nay, it may be said rather to have retrograded; else we should not consider those productions of antiquity which time has spared to us, as fit models for our present imitation. That science, on the other hand, participates in this mutability, no one would think of denying; but that it is not equally affected with art is very manifest. Before the invention of printing, indeed, there was good reason to apprehend, that the world might lose the knowledge acquired by its sages: but the discovery of that noble art has given to the true philosopher a channel of permanent communication, with succeeding ages; he can bequeath to posterity, in a compendious form, those truths which have resulted from a life of study; and he can enable those, who wish to tread the path which he is quitting, to start from the point at which his enquiries terminated: so far as his discoveries extend, and so far as his deductions therefrom are sound, so far are his works imperishable, because they relate to things which are, in this world at least, unchanging. Had the ancients busied themselves with the study of comparative anatomy, and bestowed upon the construction of the common animals of their country, one half of the attention and talent that was lavished upon other studies, their writings on natural history would be just as valuable now, as they would have been then; and the works of Pliny, instead of being a tissue of fables and absurdities, would have held the same rank with us as those of a Savigny or a Cuvier.

Mutability in science only belongs to *error:* for truth, no less than nature, is unchanging; whereas mutability, on the contrary, is a necessary accompaniment of art, and is interwoven with its very excellence.

(48.) There is an inexpressible satisfaction, an intellectual delight, in the pursuit of truth, which few but the philosopher can fully understand. This luxury of the soul, as it may well be termed, belongs more especially to the pursuit of natural science; particularly to those branches which are usually termed demonstrative. The man who studies the forms of nature, has before him, so far as those forms are concerned, models of perfection. He has no need to suspect that others exist, in distant countries, more perfect of their kind, than those before him, and which he should previously see and study. He has not to consult popular taste, ephemeral fashion, or arbitrary opinion, on the value or importance of his pursuits. He has before him *truth:* his sole business is to analyse all the parts and all the bearings of that truth, and make them known to the world. The models and materials of his study are divine; and how much they exceed those of any human artist, will be manifested by a blade of grass, compared with which the most exquisite carvings in stone or ivory sink into insignificance. The calculations of the astronomer, and the results of the chemist, are productive of much the same feelings. Truth indeed is but seldom attained, yet with superior minds this very difficulty serves but to increase the ardour of its pursuit.

(49.) Another advantage, almost exclusively belonging to the natural sciences, is this, that they carry the mind from the thing made, to Him who made it. If we contemplate a beautiful painting or an intricate piece of mechanism, we naturally are led to admire the artist who produced them, to regard his superiority with respect, and to enquire who and what he is. We mention his name with honour, and take every fitting opportunity of extolling his talents. If such are the effects of contemplating human excellency, how much stronger will be the same train of thought and of feeling in the breast of every good man, when he looks into the wonders of the natural world, and thinks upon the surprising phenomena which it exhibits! When he sees that this globe is inhabited by incalculable millions of living beings, all different from himself, his pride will be humbled by this conviction, that the earth was not made for him alone. And when he finds that all these beings, however minute, or, to the vulgar eye, contemptible, have their allotted station and hold their distant course in the great operations of the universe, he is led to enquire into his own nature, and to look towards that Great First Cause, whose bounty created, and whose providence sustains, such hosts of creatures. Those pursuits, in short, which are most calculated to expand and elevate the mind, are unquestionably the most noble; and none can be ranked above those which lead us to contemplate the Deity; to look, in fact, from the effect to the cause; and to be impressed with enlarged notions of that stupendous power and ineffable goodness, which pervades all matter and all space.

(50.) Such are the most striking advantages, in

reference to the human mind, resulting from the study of the natural sciences, generally so termed; but there are some which more especially belong to natural history, and which are not unworthy of a more particular notice.

(51.) Before, however, we proceed farther, it seems desirable to explain the real objects of the science we are now engaged upon. What, therefore, are the truths it is intended to teach? and what are those deductions it is calculated to unfold? In giving the following definition of natural history, we think far preferable to pass over, *sub silentio*, the vague or the erroneous opinions of others; since our object is not to lay before the general reader controversial arguments, or to embarrass the student by contrariety of opinions.

(52.) The object, then, of natural history is, to make known the different animals, plants, and minerals existing on the earth, in such language, and with such precision, as will enable them to be recognised by those who study. This is the general scope of the science; but it more properly comprehends three distinct objects of enquiry, by attending to which the nature of the whole will be better understood. If we consider in what manner any object in nature can be most effectually made known, we shall find that this knowledge embraces the consideration of the following particulars: —

1st, An examination of its individual structure, both internal and external.

2dly, A history of its economy: and,

3dly, The determination of its rank or station in the scheme of nature.

Whatever belongs to the history of an animal or a plant, is comprehended under one or other of these heads of enquiry. The definition, however, is not so applicable to mineralogy, inasmuch as inanimate objects cannot be said to possess either habits or economy. If, however, we substitute for these properties the growth or production of minerals, generally so called, the above exposition will be applicable to all the three kingdoms of nature.

(53.) Now, in estimating the measure of labour or of talents necessary to the successful prosecution of these several objects, we perceive that they are suited to different degrees of intellect; and, consequently, that there are departments which can be prosecuted, with advantage to the whole, by men of moderate ability and limited information. So wide, indeed, is the scope which this science embraces, so multifarious are the points of information to be elicited, and so easily may many of these points, under peculiar circumstances, be elucidated, that there is room for the beneficial labours of the youngest student, no less than of the most matured and philosophic mind. The successful prosecution of natural history, like that of all other demonstrative sciences, depends upon facts; and when we consider the number of the data necessary to complete the history of an individual species, and then reflect on the hundreds of thousands of species which exist upon the earth, we shall immediately perceive that every attentive observer has the power of contributing something towards his favourite science; something which has been yet unobserved, or, if observed, unrecorded. He may thus remove the veil from one stone at least of the temple of nature; or he may, by the discovery of one single but important fact, clear away an accumulation of doubts and difficulties that have long impeded the path of the greatest adepts. Let us not, therefore, affect to despise, as some among us have done, the describer of species; but remember that in the temple of nature there are niches for all her votaries.

(54.) Natural history has generally been termed a science of observation; and such, in a restricted sense, it undoubtedly is. The error of the definition is this, not that it is untrue, but that it is partial and insufficient. What would be thought of an astronomer who defined the study of the heavenly bodies to consist in a correct nomenclature of the stars, an accurate computation of their relative magnitudes, and of the various appearances which, under particular circumstances, they assume? Suppose, also, that the business of the mineralogist was simply to study the external forms of the substances of the earth, to compile a dictionary of their names, and to point out the uses to which they could be applied. In either of these cases it would be manifest that the essential philosophy of these sciences would be lost sight of; that we should merely be regarding the surface of things, and be busying ourselves about effects, to the utter neglect of those great and sublime causes which are unfolded by the laws of gravitation, the theory of motion, and all those splendid truths which give such dignity to these sciences. As it is with astronomy and chemistry, so is it with natural history: knowledge of individuals, and of the facts which belong to them, is undoubtedly the basis upon which this and all other sciences repose: but if the zoologist or the botanist contents himself with this information, — if he remains satisfied with isolated descriptions drawn up in technical language, — and compiles a dictionary, under the name of a system, of hard names, he has no more right to term his pursuits intellectual, or to dignify them with the name of science, than the astronomer would have, under the circumstances just supposed. All branches of natural science, however varied may be their materials, or however diversified their nature, have but one and the same object in view — the discovery of the primary laws of nature. In comparison with this, all other objects, however superior they may be in point of utility, yet, in reference to sound philosophy, are of a secondary or subordinate nature. As all sciences are based upon facts, known, or to be known from experience, so are they, in their early state of developement, matters of pure observation. It is only when we have acquired the power of generalising these facts, when such generalisations agree among themselves and with every thing we see or know of nature, that the theory of a science becomes either absolutely demonstrative, or approaches so near to certainty, by the force of analogical reasoning, that it is not contradicted by any thing known. The case of natural history, then, is precisely this: in its early stages it is a science of observation; in its latter, it is one of demonstration. There are few, indeed, even among philosophers, who have the least suspicion that natural history is deserving of this character.

But the question resolves itself into this, Are there any fixed and universal laws by which the variations of the forms of nature are regulated? If this question can be answered in the affirmative; if all these variations can be traced to certain primary types, following each other in one constant and unchanging series, we have the most conclusive evidence that human research can elicit. It will be our especial object, therefore, in the subsequent volumes of this work, to demonstrate the truth of this proposition, appealing for its stability to those facts with which we first commence the fabric of the science, and which, coming within the range of ordinary observation, it will be in the power of every one to verify or disprove.

CHAP. II.

NATURAL HISTORY VIEWED IN ITS CONNECTION WITH RELIGION. — AS A RECREATION. — AS AFFECTING COMMERCE AND THE ECONOMIC PURPOSES OF LIFE. — AS IMPORTANT TO TRAVELLERS.

(55.) I. THE nature and objects of the science having now been sufficiently explained, we may consider the advantages which more peculiarly attend its prosecution, independent of those which have already been noticed, in a general way, as belonging to all intellectual pursuits. We shall enumerate these advantages under distinct heads, because some are applicable only to particular persons, objects, or circumstances; and because, by so doing, we may excite an interest and a love for these enquiries in the minds of many persons, who imagine they have neither the abilities to study, nor the means of adopting such pursuits, and of others who think they are in no way interested in them. We shall therefore look to natural history — 1. as connected with religion; 2. as a recreation; 3. as affecting the arts and common purposes of life; and, 4. as an essential accomplishment to the traveller.

(56.) All the advantages that result from science, are comprehended under two distinct classes: — 1. Either they relate to our worldly prosperity, by opening new sources of wealth, of convenience, or of luxury; or, 2. they administer to intellectual gratification and our spiritual welfare. When, therefore, we speak of the advantages attending the prosecution of this science, we must readily admit that they chiefly belong to the latter class, although they may, in a limited degree, be applied to the former. The great characteristic, however, of natural history, is its tendency to impress the mind with the truths of religion; and thereby of improving and regulating the moral feelings. Its application to the wants of man is comparatively slight, and generally so remote as not to be immediately perceptible. It has not, like chemistry, been employed to the improvement of manufactures, nor can it contend with botany in adding to the luxuries of the table or the elegances of taste. It very rarely opens a new source of commerce, nor can it assist astronomy in giving power and confidence to the mariner. Neither does it lead, like other kindred pursuits, to pecuniary advantage, public employment, or academic honours. Natural history, therefore, will never assume its real station in a commercial country like this, so long as it is not protected and fostered, encouraged and rewarded, by the government. The office of natural history is to expound the works of Omnipotence; and it becomes, from that very circumstance, one of the most dignified that can employ the human mind. It seems, in fact, to be that peculiar study which is, above all others, most designed to bring man into communion with his Maker. In this respect it is even superior to astronomy. The grandeur of the heavenly bodies may speak more impressively to our senses, and their periodical movements excite, at the moment, a greater degree of wonder; but all enquiry into their precise nature is futile. We know not whether those distant worlds are inhabited by mortals or by spirits; whether they are the abodes of imperfect beings like ourselves, or of spirits exempt from sin. All this is hidden from human research. But with natural history the case is different: the objects of which it treats are continually before us: we can, in a great measure, distinguish their properties, examine their structure, and explore their economy: the most minute parts of their organisation can be investigated, every nerve traced, and every substance analysed. And if our knowledge of the system upon which they are formed, has hitherto borne no comparison with that which we have acquired in other physical sciences, it is only because the minds of men have dwelt upon minute details, instead of searching for universal principles.

(57.) It may be thought unnecessary, perhaps, in a work of this nature, to advert to those reflections which arise in a religious mind, on contemplating the works of nature, and which, upon some occasions must force themselves on the notice of the mere worldling. One of the first impressions which arises on studying natural history, but more particularly animals, is, the conviction of *design* in their creation. And this design not only relates to the formation of an animal to effect a particular purpose, but is equally manifest in the peculiarity of its structure, the season when it is most active, and the means by which it effects its allotted object. The moment we arrive at the conviction of *design* in the material world, we are persuaded that there is a *Designer;* or, in other words, the atheistical doctrines of chance, and of self-development, vanish like a mist. This design must have emanated solely from the Creator; and as *perfection* is His attribute, *design can never be partial, because it would then be imperfect.* Every thing in nature being thus formed for some specific purpose, it follows, man was created with the same object. But what this object is, unassisted reason can never discover. It requires no depth of penetration to perceive that one of the chief uses of the vegetable kingdom is, to supply food to the animal; this object being effected, the plant dies. Insects either furnish nourishment to other animals, or they assist the propagation of plants, or they hasten the decomposition of decayed matter; this done, the purposes of their creation appear to be effected, and they pass away. In like manner we may trace the great outlines of *design* through every branch of the animal kingdom : each is dependent the one upon the other, and this dependence produces the most inconceivable harmony.

(58.) But when we come to MAN, who reigns over the whole, — when we ask for what visible purpose, or with what design, *he* was called into being, our natural reason is baffled. No part of the economy of nature is dependent upon his existence: he assists not in one of the innumerable operations which are continually going on, by which the harmony of nature is upheld, and a mutual dependence preserved in all the parts. The fruits of the earth require not his care, nor do the beasts of the field need his protection. His power is not wanted to prevent the increase of noxious animals; for his Creator has chosen other and more humble instruments to effect such an ignoble purpose. The rapacious tribes of quadrupeds, of birds, and of insects, keep their respective classes within due limits, while it has been ordained that these animal destroyers should propagate slowly and sparingly We find, moreover, that, in countries very thinly inhabited, there is no disproportion between those animals which are predacious, and such as live upon vegetables. Man, in short, although the noblest work of nature, is yet so unnecessary to her operations, and so disconnected with all those designs she is carrying on in the material world, that his absence from the earth would not be missed. He rather impedes than advances the free developement of her works. In this point of view he is inferior to the very worm he treads upon; the extermination of whose race would render the earth unfruitful, and bring famine and death upon its inhabitants. It may be argued, indeed, that the design of the Creator, in calling into existence this last and best of his works, was to give him happiness, to fill him with delight at the wonders which surrounded him, and that he should do good to such of his creatures as he was to govern. But had he been created solely for those purposes, we should have seen them accomplished; because imperfection in the means for accomplishing the end belongs not to the Omnipotent Being. What, in short, do we actually see? Human happiness is a shadow. The mass of mankind are totally indifferent to the wonders of creation; and cruelty to the beasts of the field is to them an amusement. Seeing, therefore, that unassisted reason is totally incompetent to solve this momentous question, we are naturally led to enquire into the truths of religion, to see whether they will explain this apparent anomaly. Here, then, we find every difficulty solved, and every doubt removed. Man discovers that the chief design of his creation is, that he should enjoy an immortal happiness in a higher region; and that he is placed upon this earth, not as necessary to its well-being, or to perform a part in its regulation, but as one who is undergoing a state of probation; who is journeying, indeed, as a stranger and a pilgrim; but who is provided with those means and aided by that assistance which may finally secure the great, the glorious designs of his Maker.

(59.) It may be questioned whether the above train of reasoning, agreeable alike to logical deduction and to indisputable fact, could thoroughly be entered upon by any one who was not a naturalist, or, at least, who had not an intimate acquaintance with some of the most remarkable phenomena of the animal kingdom. Hence it is manifest how intimately the study of nature is connected with the truths of religion. Every philosophic argument which can be drawn from the material world, in corroboration of the books of Scripture, will tend to bring those who doubt, to investigate their pages more closely; while those who already believe their divine inspiration, will have that belief strengthened and confirmed, rejoicing that sound philosophy bears witness to those truths which they feel to be immutable.

INFERENCES FROM DESIGN.

(60.) Such are the evident conclusions which result from a conviction of design in the creation. And this conviction will be equally attained, whether we take an enlarged view of the subject, or descend to minutiæ: whether, with the scholar or the philosopher, we discuss the question by the rules of logic; or whether, with the ordinary observer, we adopt the more simple process of contemplating those innumerable and beautiful objects of the creation which lie before us. If every thing in nature which we examine and reason upon, evinces this principle of *design*, it follows that design is universal (57.). And as experience teaches us, that, although we can trace the principle, we know but a limited portion of its extent, it may be fairly inferred that even of that portion which man may discover, we know as yet but an insignificant part — and that, too, is seen " as in a glass, darkly." How little, for instance, do we know of the manners and instincts of the common animals around us! and how little have we yet learned of the purposes for which they were created! Now, as the Author of this principle of design is Himself the type of perfection, that perfection must extend to all His attributes. Hence arises the supposition, that every created thing has a twofold use; one in relation to the economy of nature, and another to the exemplification of moral and religious truths The first is palpable to the most illiterate observer: every one, for instance, can see, that without insects, there would be no occasion for spiders; and that without swallows, we should suffer from a plague of flies. But the moral use of the book of nature is not so apparent. We can, indeed, perceive how forcibly, though silently, the duties of industry, perseverance, order, and subordination are exemplified in the ant and the bee. Yet, if this was the only moral or religious precept that could be learned from the study of nature, we might be tempted to think the application of this science to moral truths was but slight; and to spiritual, no greater than that of proclaiming the existence and the perfection of their Creator.

(61.) That there is a general analogy between the different parts of the animal world, by which one object or group represents another, is a truth so universally admitted in modern science, that it need not be here advocated. It is confirmed, not only by the most profound investigations, but is perceived and assented to by the vulgar, who, in many instances, have given to particular animals such names as express an intuitive perception of this principle, without the power of demonstrating the analogy implied by such epithets. Hence the origin of such names as night *hawk* and Tern *owl*, as given to the goatsuckers; *chauve souris*, or flying mice, applied by the French to the bats; water *hens*, to the *Fulicæ;* sea *swallow*, to the Terns; and *swallow* butterflies, to the genus *Podalirius*. The provincial or vulgar names of well-known animals, in every language, furnish innumerable instances of the same perception of natural analogies. These resemblances, therefore, being undeniable, we must come to one or other of the following conclusions:—Are we to consider them as partial and incidental, incapable of being reduced to any definite rules, and governed by no fixed principles? or, are we to view them as the prominent features of some part of the plan of creation; as the strong indications of something *beyond* the surface of things, and as forming a portion of some great *system* of harmonious relationship? Upon this point, again, the scientific world has been set at rest. The theoretical inference which would favour the last of these suppositions, has been demonstrated to be correct both by reason and experience; and we now know that all these resemblances are to be traced to one universal and consistent plan, as similar in its laws, as it is harmonious in its results. Here, then, is opened an exalted and a boundless field of design; wherein the Christian philosopher is not only enabled to draw proofs of the Divinity from the individual objects, but from the system by which this endless diversity of forms is regulated.

(62.) The results attending the investigation of this system of representation, having been uniform in every department of nature yet investigated, we are led to enquire, what further can be learned?— whether there be still any ulterior design, for the instruction of man, beyond those which we have discovered? and whether the knowledge thus gained by analysis, can be applied to the illustration of higher truths connected with our spiritual welfare? On this point, again, the Christian philosopher will have no doubts. He is told in that inspired volume in which *he* at least believes, that " we see now, as it were in a mirror, the glory of God reflected *enigmatically* by the things that he has made."*
He is thus assured that the book of nature

---

* 1 Cor. xiii. 12.   See also North. Zool. Introd. ii. p. lvi.

is a book of symbols; and if he require further evidence of this assurance, he finds it in the concurrent opinions of some of the greatest and most learned men whom the world has produced. The existence of an analogy between the material and the immaterial world has been a doctrine of firm belief in all Christian ages, and has been illustrated with force and eloquence by many powerful writers who were not men of science. There is one, however, now among us who unites in himself the pious divine and the scientific naturalist, whose words are too remarkable not to be here quoted. "The instruction of man was best secured by placing before him a book of emblems or symbols, in which one thing might represent another. If he was informed by his Creator that the works of creation constituted such a book, by the right interpretation of which he might arrive at spiritual verities, as well as natural knowledge; curiosity, and the desire of information concerning these high and important subjects, would stimulate him to the study of the mystic volume placed before him; in the progress of which he would doubtless be assisted by that divine guidance which even now is with those who honestly seek the truth. Both divines and philosophers have embraced this opinion, which is built upon the word of God itself."—*Introduction to Entomology*, vol. iv. p. 402.

(63.) From the doctrines of affinity and analogy, which will subsequently be fully discussed, we learn two great truths. First, that the progression of the affinities of nature is circular; that is, every natural group has its objects disposed in a revolving series, so that the *last* joins to the *first*, as well as to that by which it was preceded. Secondly, that three of these circles always unite among themselves, and form a larger circle. Now these laws, it must be remembered, repose upon the firmest of all foundations, namely, that of analysis; and are, consequently, capable of demonstrative proof. When, therefore, we find these laws hold good in every division of the animal world — when we discover that the contents of one circle are represented by those of another, and that by no other theory can we explain those innumerable phenomena and relations which we see in nature — we cannot for a moment believe that this extraordinary harmony is not a part of the system of creation.*

(64.) II. If we are asked, what are the chief uses and what the advantages of natural history? we should reply, that it not only leads us to look to heaven, but that it opens one of the greatest sources of happiness on earth. In the preceding pages, we have dwelt sufficiently on the first of these topics; we will now enquire into the second. It might be expected, perhaps, that, before we enumerated these minor advantages of natural history, we should show in what manner it is an intellectual science; and thereby make good its claim to be ranked among those which, for their successful

---

* The Christian philosopher will not fail to perceive the interesting field of enquiry which here expands itself; in which he may observe the close analogy that exists between the revealed character of God, and the material creatures of His creation.

prosecution, require the higher energies of the human mind. That it is truly of this description, might be readily inferred; for the works of nature are much more difficult to understand than those of man. Yet, did we at once expatiate on the deep research necessary to acquire proficiency,—did we detail the many and varied acquirements essential to a high cultivation of the science,—we might possibly frighten those away, of whom we should otherwise have made disciples. We prefer, therefore, a more agreeable and a more inviting course. As our Series of discourses upon Nature is intended to be elementary, we shall commence from the lowest step, and gradually ascending, conduct the student from the leading principles of all sound knowledge in this department, to an acquaintance with its details. Let us now, therefore, regard natural history rather as a recreation than as a science,—as a pursuit for the man of leisure, and a relaxation for the man of business: we will consider it also as the means of acquiring and preserving health, and as a source of pleasure to the valetudinarian.

(65.) The study of Nature can never be so well or so delightfully prosecuted as in her own haunts, "remote from cities." Hence it is, that no pursuit can be better adapted for a country life. We are then, as it were, in the boundless temple of Nature, and we explore her truths in all its various recesses. The tediousness of a country life is proverbial; but did we ever hear this complaint from a naturalist?— Never. Every man who in his walks derives interest from the works of creation—who looks to the habits, the instincts, and the forms of animals—and who reflects upon what he sees, — is, in spirit, both a naturalist and a philosopher. To him, every season of the year is doubly interesting; for, independent of those changes apparent to all, there are others which bring peculiar delight to himself. With each succeeding month, new races of animals and plants rise into existence, and become new objects for his research: these, in their turn, pass away, and are succeeded by others; until autumn fades into winter, and both the animal and the vegetable world sink into repose. But even this ungenial season, so dreary and comfortless to the mere country resident, is not without interest to the naturalist; for no period of the year is so unsuited to animal life, as to leave our climate destitute of inhabitants. A fine sunny day, in the depth of winter, calls forth many little insects, rarely seen at any other period: while the numerous mosses and lichens, then in fructification, give to the woodland walk of the botanist a new and lively interest. Nor are the naturalist's pursuits suspended when storms prevent his walks, and confine him to the house. The acquisitions to his various collections, made during the past year, are to be examined and arranged; or his loose notes are to be compared and digested. These are fit and delightful occupations for the long winter evenings; and over a cheerful fire, he only laments that the hours glide too rapidly away.

> "Thus may our lives, exempt from public toil,
> Find tongues in trees, books in the running brooks,
> Sermons in stones, and good in every thing."

(66.) The enthusiasm of naturalists is very apt to

surprise ordinary people; but it may be explained on very simple principles. Every one, raised above the condition of a clown, is in a greater or less degree sensible of the beauties of nature, as seen in a fine landscape; but on none do such scenes make a stronger impression than upon the painter. He, and he alone, is able to analyse, as it were, the picture before him, and to understand *how* that general effect, which is merely judged of by others as a *whole*, is produced in detail. By being able to do this, he feels the beauty of picturesque scenery in a much superior degree to others. The same feelings influence the naturalist: he walks abroad with others, and admires with them the *general* beauties of nature, but his perceptions of them are keener, because he understands them better. A thousand little circumstances, unobserved by ordinary eyes, attract his attention, and call forth his delight: the plants, the birds, or the "creeping things," that he meets with, are known to him by name; he understands something of their modes of life; they come before him as old acquaintances, or, if as new ones, they are doubly welcome. While his companions are wondering, and enquiring of each other the name of a beautiful flower, a curious insect, or an uncommon bird, he is seldom at a loss for a reply. He is, in fact, conversant with those things before him, which are strangers to his companions. And as we always feel pleasure in proportion as we understand that which produces it, so does this feeling frequently rise to enthusiasm both with the painter and the naturalist. When these two pursuits, indeed, are united, we can hardly imagine a higher degree of intellectual gratification than that which they afford.

(67.) The amusements of the country are generally expensive. Field sports cannot be followed without horses, and dogs, and guns; and these lead us not unfrequently into the society of men with whom we have no other feelings in common. But the quiet student of nature has no need of such paraphernalia. the few implements of his chase are easily and cheaply procured ; nor is he called to celebrate his feats over deep potations, or to make them the subject of boisterous mirth : his pleasures are intellectual, and therefore tranquil. Seldom, indeed, does he meet, if far removed from towns, with companions like himself, with whom, at the close of day, he can talk over its events ; but, if he be a man of leisure, occasional intercourse with such congenial spirits can generally be accomplished. Short excursions, even for a day, may be compassed, even by the most busy. A new district may be resorted to, and explored. Similarity of pursuits not only elicits information, but animates zeal ; and we return to our solitary walks with renewed vigour. Nor are the pursuits of the country naturalist altogether inapplicable to practical uses. The various injuries which affect the produce of his garden or his fields, call for his investigation, and may frequently be remedied by his care. How much damage, for instance, is annually done to our fruit-trees by their insect-enemies, few or none of which are thoroughly understood. We scarcely know a publication which would be more useful, or more generally popular, than one which should be devoted to the history of such insects as are injurious to fruit and timber trees; and none but a country naturalist could write such a book. Scientific learning is not essential to the undertaking, seeing that the insects themselves, if thoroughly well described, could always be named or identified. The same enquiries, directed to those insects which infest our grain or other agricultural produce, as hops, turnips, &c., would be still more beneficial, and might be the cause of preventing, in some instances, great loss, if not total ruin, to individuals. Let it not be said, therefore, that the pursuits of the country gentleman, who may be attached to natural history, are either trivial or unproductive of real benefit. They embrace, in fact, the investigation of those subjects, which render natural history subservient to the economic purposes of life. And if ever the agricultural world is enlightened on these matters, the information must come from those who study nature in the fields and woods—not in libraries and museums.

(68.) To the man of business, confined during the day to the closeness of an office, or harassed by the anxieties of his profession, relaxation is always welcome; but it becomes doubly so, when the mind is at the same time instructed and delighted by pleasing images. Those who are engaged in business, cannot always enjoy the recreations of the country naturalist, or gratify their love of nature by contemplating her works in the fields: but no situation precludes the use of books, or the formation of collections. Next to the actual sight of foreign countries, and the study of their living productions, nothing brings them before us so completely as the narratives of travelled naturalists; of those who have personally explored the various regions of the earth for the love of natural history, and who bring before us the manners and peculiarities of animals in a state of liberty. What a love for such pursuits is inspired by the animated and poetic pages of Wilson! and while we read the Northern Zoology * of the late adventurous naturalists to the Arctic regions, we feel almost prepared to encounter the same difficulties, for the sake of participating in the pleasures of the journey. Among the numerous series of pocket volumes published in these days, and adapted to all the different branches of entertaining knowledge, we should like to see the announcement of one which embraced the travels of naturalists. No collection, if judiciously made, would be more permanently and generally interesting to the lover of nature, whether he pursues natural history as a study or as a recreation.

(69.) We shall subsequently show that some knowledge of the productions of other countries may furnish beneficial hints to the merchant in his foreign speculations; and that he may, in this instance, turn even his recreation to a profitable advantage. If he is prevented from collecting natural productions himself, they can always be purchased; and at no time so cheaply as at present. A few pounds, judiciously expended on proper objects, will be quite sufficient to procure an elementary collection of

---

* The title of this work, published under the authority of the Government, is *Fauna Boreali-Americana;* but, for the sake of brevity, I generally quote it as *Northern Zoology.*

birds, insects, or shells; in the arrangement and study of which, assisted by a few elementary books, he will find a mental and fascinating recreation, far exceeding that derived from the glare and suffocation of a theatre, or even from the levelling monotony of cards.

(70.) There is a quietness and a placidity in all that relates to nature, which is particularly congenial to the spirit of a good man, and which renders his pleasures independent of the auxiliary aid of the world. They are beyond the influence even of fashion; they do not, necessarily, bring with them contentions for superiority, the murmurings of envy, or the miseries of disappointment. The true naturalist loves science for her own worth,—for her own dignity. He quits the haunts of folly and of idleness, for his study: there, in converse with a friend of kindred spirit, or, if blessed with a family, with those of his own circle, he enjoys the pure delight of receiving or of imparting knowledge. There is always some new fact to be imparted, some new book to be talked of, or some new acquisition to be shown and admired. The man of business wants relaxation; but when that is sought for in the excitement of mixed society, or of public amusements, diversion *may*, perhaps, be found, but repose cannot. The man, whose profession keeps him in the bustling scenes of life the greater part of the day, must choose his recreations either abroad or at home. By all but the gay and giddy, who have yet to learn what the world really is, the latter resort will be preferred. But retired or domestic life does not necessarily suppose idleness, and the cultivated mind will not even then be satisfied with mere commonplace conversation. Home can only be truly enjoyed, where a taste for some one rational pursuit exists; and among these, there are few which promise more delight, than the love of natural history.

(71.) Our science is no less conducive to health, than to rational pleasure. It requires to be prosecuted by different means — all tending, indeed, to the same point, yet carried on by different individuals, under different circumstances. The practical and the closet naturalist have each their respective departments, equally essential to the advancement of science, although very different in their duties. Facts regarding habits and instincts must be sought for in the fields and the woods; while their application and generalisation can best be meditated upon in the closet. Exercise is essential to health; and a lover of nature wants no other inducement to secure such a blessing, than the active pursuit of her treasures. It is curious to remark the great age which naturalists generally attain. Whether this longevity is to be attributed to those quiet and temperate habits inseparable from their studies, or to that exercise necessary to active investigation, certain it is, that both must have considerable influence on the prolongation of life. An entomologist, having no professional occupation, and ardently devoted to his study, may be said to live, during the greatest part of the year, in the open air. No soil or situation is unproductive of *his* game; he has not to wait until the First of September, for free licence and permission to capture. No sooner do the first mild beams of a spring sun awaken the insect world into life and motion, than he prepares his tackle, and commences sport. His exercise is attended with a combination of pleasures. He quits the beaten path and the dusty road, and wanders, as fancy leads, " through woods, and lanes, and coppice green." He admires nature as a whole, as well as in detail. He reposes, in the heat of the day, beneath the shade; and returning to his frugal board, refreshed in mind, and invigorated by health, partakes of what is spread before him with a relish and an enjoyment unknown to the indolent. It is delightful to read with what enthusiasm the amiable and excellent author of the *Lepidoptera Britannica* speaks of his youthful entomological excursions. " I have diligently examined," says Mr. Haworth, " many parts of England personally, and usually on foot and alone; but sometimes accompanied by pedestrian friends of congenial sentiments and taste. Industriously have we sought, and never once in vain, a great variety of woods and lawns, hills and vales, marshes and fens; one summer only, travelling, in various journeys, not fewer than a thousand miles, in spite of heat and cold, wet and drought, and other concomitant impediments." (*Lep. Brit.* Preface, x.)

(72.) How frequently do we hear valetudinarians express a repugnance to exercise, particularly in country situations, because they have no object to take them abroad! They are obliged, forsooth, to walk for the mere sake of walking; while all those pleasurable feelings, which physicians tell us are so essential to the full benefit of exercise, are destroyed by the consciousness of performing a task. Could such persons once enjoy the pleasure experienced by the field naturalist, they would no longer complain. The hedges which might be seen from their windows would furnish subjects for research; and they would require no other object than to ascertain by what races of the insect world their own neighbourhood was inhabited, what plants grew in their fields, or what birds visited their trees. The smallest inclination towards such tastes would beget the taste itself: regular and daily exercise would powerfully aid the return of health, and pleasurable occupation would produce serenity of mind.

(73.) With all these concomitants, there are few invalids, except the infirm, the aged, or the diseased, who would long remain so. But even those who are physically incapacitated from sharing in the active prosecution of natural history, may still derive, from its passive pursuit (if we may be allowed the term), a never-failing source of rational pleasure, if not of mental study. If they cannot collect, themselves, they can send others to do so; and if foreign productions are required, the commercial naturalists of London are continually receiving new objects, from which selections may be made. An amiable and highly accomplished female friend, whose name, on other occasions, we have more than once mentioned, during a long and protracted illness, occupied herself in forming a beautiful *Hortus Siccus* of our native plants. An intelligent servant was the *active* collector; who, without any knowledge of botany, brought to her mistress all such plants of the neighbourhood as were not absolutely common weeds. Seated in her arm-chair, our enthusiastic friend could manage to select and dry such as she wanted, and occasionally to examine those that were new to her. If I remember right, not days or months, but even years, passed in this way. Botany and conchology relieved the wearisomeness of reading, and gave to her long period of sickness a degree of relief perfectly inconceivable to those who possess no such resources.

(74.) An anecdote of a late noble and munificent patron of natural history — Sir Joseph Banks — well illustrates what we are now recommending. When that enterprising naturalist, leaving the comforts and the luxuries of wealth, embarked with Solander to share the dangers and privations of a circumnavigating voyage,—arrived at Rio de Janeiro, the jealousy of the Portuguese authorities was so great that not one of the party was permitted to land. This prohibition must have been excessively mortifying to all; but how much more so to Sir Joseph and his companion, who beheld from the deck a noble and richly wooded country, covered with tropical vegetation, and abounding in unknown plants! But the celebrated botanists did not despair. Having taken in some live stock, and having still one or two sheep and goats, they were permitted to receive fresh fodder every day from the shore. No sooner did it come on board, than Sir Joseph and the Doctor began their herborisings: the bundles of grass and herbs were diligently examined, and many new plants were found, either in flower or in seed; the former were carefully dried, and many of the latter subsequently vegetated in the hothouses of England. Pecuniary reward induced these botanical purveyors to bring on board bundles of other plants; so that, confined to the ship, and incapable of procuring a simple specimen by their own exertion, they were yet enabled, by this simple yet ingenious method, to secure a comparatively large collection of Brazilian plants. I may be pardoned, perhaps, for adverting to my own experience on this subject. While exploring that almost interminable line of virgin forests which run parallel with the coast of the Capitancies of Bahia and Pernambuco, I was attacked by a cutaneous disease of the country, and incapacitated from walking. Yet this mortification, great as it was, caused but a partial suspension of my zoological researches. Two of the Indians who accompanied me had been trained as entomologists, and another was a crafty hunter. All three were despatched every morning in different directions, and in the evening returned with their zoological spoils. Never shall I forget with what exquisite sensations of anticipating hope I watched the declension of the sun, and looked to the return of my people, for whom a warm supper had been partially prepared by myself. Every day brought something new to my collections, and provided us with food for the next: the morning was spent in preparing the skins of birds, and finally arranging the insects; while the evening was occupied in examining and rejoicing over new acquisitions. It was only in the intervals which occurred between the accomplishment of the one, and the near approach of the other, that the spirits were sometimes depressed, and the ailments of the body severely felt.

(75.) Natural history has this peculiar advantage — that it can be prosecuted, in one shape or other, by almost every body, and under every ordinary circumstance. Of all sciences, it is that which requires, in most of its departments, the fewest materials. It is as much within reach of the cottager as of the professor; or rather, we should say, it embraces questions which can be solved by the former, just as well, and frequently much better, than by the latter. If, as is generally the case, the amateur confines his attention to the productions of his own country, three or four elementary books, and as many implements of chase, are all that is requisite. His own exertions will procure him a collection; and he thus furnishes himself with additional materials for study: but even these are not absolutely essential. The appearances of nature can be investigated and recorded without acquiring the technicality of scientific language; nor, for such purposes, are collections or museums indispensable. A fund of interesting anecdotes of our native animals may be collected by an attentive observer, who is nevertheless ignorant of their scientific names. White, of Selborne, is a striking example of this truth. His letters show a very confined knowledge even of the imperfect arrangements of the period in which he wrote: yet how delightful are his observations! The fact is, he looked to nature, and simply recorded what he saw. The writings of such men are invaluable, because, while systems change, nature continues the same. The recently published *Journal of a Naturalist* may be taken as a model for such remarks: both works show that the subject is inexhaustible; and both may teach all who live in the country, what sources of rational pleasure are within their reach, by merely looking to the productions of their own neighbourhood.

(76.) In tracing thus far the advantages of natural history, the recorded opinions of others have been confirmed by our own experience. But there remains one period of our existence, at which its effects upon our mind can at present be only imagined, although we humbly trust we may have the power of confirming our present belief from experience. We allude to the feelings that result from such pursuits when old age comes upon us; and when we naturally look back to the route we have chosen for the journey of life. That our present ardour will subside, we can well imagine; but we believe that it will never degenerate into indifference. It has, indeed, been mercifully ordered by Providence, that our interest in temporal things should progressively diminish, in proportion as our time draws near for quitting them. But if our recreations have been innocent, and our pursuits intellectual, they cannot, in the nature of things, leave behind them regret or disappointment — much less can they inflict remorse. We can imagine, therefore, that the old age of a true naturalist, — one who looks from the created to the Creator, — must be peculiarly happy. He may have had his share of the sorrows and disappointments incidental to mortality; but they have neither originated in the sensuality or intemperance of his amusements, nor in the ambitious nature of his pursuits. Neither wealth, nor titles, nor honours, have ever had the power to lure him from his peaceful studies; and he is, therefore, exempt from the committal of those mean artifices and unworthy acts, by which such distinctions are too often gained. We can imagine such a man looking back on the quiet path he has trodden, with something of the same feeling with which we contemplate, from a mossy seat, the vista of a green embowered lane, nigh to which is the public road, sultry and dusty, thronged with crossing vehicles and jostling crowds. Although no longer fit for active exertion, we can still fancy him contemplating his collections—the acquisitions of his youth, and the study of his manhood—with that complacency which we feel towards an old companion. Every object in his little museum has its own story; the scenes and incidents of youth are brought back to his recollection in all their freshness; and the memory, dwelling on these green spots in the desert of life, will oftentimes be prevented from recalling others of a less cheering nature. He looks abroad in the spring of the year, and sees the face of nature renewed, with the same beauty and freshness, as when he contemplated her in the spring of youth. *That* season of his life has long passed away: but he knows that he, too, will be renewed—that *his* winter will be changed to an eternal spring; and with firm but humble confidence in the promises of his God, he resigns the contemplation of His sublunary works, in the sure and certain hope of seeing those which are heavenly.

(77.) III. Natural history is now to be considered in reference to commerce, and the economical purposes of life. It has always been remarked, that this study, when viewed only in reference to its economic uses, possesses few decided advantages; and that even these, for the most part, are indirect. It will not, however, be either unprofitable or uninteresting to view the subject in this light, and to enquire what benefits can result to the merchant, and the agriculturist, from acquiring some knowledge, at least, of the science now under consideration.

(78.) All commerce is derived from the productions of nature, whether in the state in which they are naturally produced, or after the raw material, as it is then termed, has been altered or worked upon by art. Every thing which administers to our wants, our comforts, or our luxuries, is derived either from the animal, the vegetable, or the mineral kingdom. It is from these great storehouses of nature that man selects such objects as he finds, by experience, are most satisfying to his wants or most adapted to his purposes. From these he derives his food and his clothing—from these he selects materials for his habitation; nor, without them, could life be supported. The vegetable and the mineral kingdoms supply us with all those medicines which alleviate pain, conquer disease, or restore health. So that, without a knowledge of the uses of those materials, life and health could not be preserved. Even knowledge itself would cease; for the pen we now hold, and the paper upon which we write, are but raw materials for communicating information,—one being taken from the animal kingdom, the other from the vegetable. Now, as one of the chief objects of natural history is to teach us the properties and uses of natural productions, it might be argued, abstractedly, that natural history is the most important and the most essential science that can be conceived; since, without being acquainted with that information which it is designed to teach, man could not exist upon the earth.

(79.) But if man, in his primeval state of rudeness or barbarism, had been compelled to study the nature of tnose things which he needed, before he had ventured to make use of them, he would have wanted both the means and the opportunity; and he might have starved in the midst of plenty. He was therefore prompted either by reason or by instinct — certainly not by science — to use those things with which nature had filled the world. He saw that certain animals, which had been destined for his use possessed tameness and docility; that they frequented his haunts, and even courted his protection. The harmlessness of their nature was apparent; and he was in this manner, probably, led to attempt their domestication, and to avail himself of their services. Hence we find, that the horse, the sheep, and the dog, were the mute companions of our primitive races. While living, their strength diminished his labour, or gave security to his property; and when dead, the greater number supplied him with food, or materials for clothing. He saw, again, that these animals fed only upon vegetables which were wholesome, and he might be thus assisted in discriminating between nutritious and poisonous roots. No information, indeed, would be more curious than that which should tell us the particular manner in which the virtues of vegetables and minerals were first discovered, and in what way it was found out that a plant naturally a deadly poison, could yet, by the expulsion of its juices, be so prepared, as to become a most nutritious food — forming the chief sustenance of nearly a fifth part of mankind.* Discoveries of this sort, however, have seldom originated in design; they have been made accidentally, generally by the uneducated savage. In proportion to the utility of the discovery, so has its knowledge been spread to others. All this, in short, had taken place, before the study of nature assumed either the name, or the intellectual character, of a science.

(80.) Seeing, therefore, that all which is essential to our wants has been already discovered, the mere superficial reader will again enquire *cui bono?* How are we to make this science practically useful? In what manner does it concern, or enter into the pursuits of, the merchant, the planter, or the agriculturist? and what good can result to them by a a knowledge of such matters? We evade not these questions, because, in the view we are now taking of natural history, they are natural and just.

(81.) Could it be shown that all those productions of nature have been made known, which possess qualities applicable to the common purposes of life,

---

* We allude to the mandioca root, from which the cassava bread of the West Indies, and the farinhia of Brazil, are made.

then, indeed, economic natural history need not be studied. But how far from this is the real fact! In regard to the qualities of animals, indeed, we cannot hope for any new discovery of importance. Though even on this point we need not despair; seeing that, but a few years ago, vaccination was unknown to us; and that it would have been deemed chimerical to assert that the cow had a property which would save millions of lives. We require not any increase to the number of our domesticated animals, for nature has bountifully made known to us all the races that we require; the horse for labour, the ox for food, and the sheep for clothing. It would be curious, indeed, though not very desirable, to see the camel, the elephant, or the reindeer, acclimated and breeding among us; but what practical good would result from this, may reasonably be questioned; while the evil of devoting tracts of ground to feed such bulky animals, at a time when the produce of our soil will not supply its human inhabitants, is sufficiently obvious. It may, in truth, be considered a fortunate circumstance for the nation, that the Zoological Society, originally formed for these very purposes, has not succeeded in a single instance, as it is said, after many years, in acclimating one race of foreign animals, either useful or ornamental.

(82.) The practical uses of natural history are not, however, restricted to such matters. A merchant who trades to a distant country must first inform himself on the nature of its productions — whether animal, vegetable, or mineral — that he may know what to send, and what he can receive. When, as

in these times, new countries are continually opening
as marts of traffic, and new channels of commerce
are making their way even into the heart of Africa,
the man who possesses this sort of information, and
turns it to advantage, not unfrequently realises a
fortune; while he, who, like the Sheffield cutler,
sent a large consignment of patent skates to Buenos
Ayres*, thinking they would, in a new country,
sell for an enormous sum, may very likely be
ruined. Every one knows the importance of our
fisheries, particularly those for the whale and the
seal. Had laws been made by our legislators for
the preservation of the former, on the same principle
as they so sedulously preserve their own game, we
should not hear of the Greenland fisheries being
almost ruined; — no one, indeed, could have drawn
up a parliamentary bill for this purpose, without a
competent knowledge of the natural history of the
animal whose race was to be preserved; — while in
regard to the seal fisheries, they might be extended,
beyond all doubt, in parts of the Southern hemisphere
hitherto entirely neglected. The fur trade, again,
opens a field for the practical use of natural history:
for, independent of the necessity of accurately discriminating the different species whose skins form an
article of commerce, how much might this trade be
extended and benefited by a merchant well acquainted
with the geographic range of these animals, the
peculiar times when their furs are in the finest condition, and what countries are destitute of such

---

* A fact which occurred in 1806.

resources! We need not insist, that such knowledge, properly and judiciously made use of, will not only be useful, but lucrative. The first traders who supplied China with the furs of America, realised large fortunes; and the same results will always attend every such enterprise, however irregular it may appear, if it is only founded on knowledge, and conducted with prudence. People go on trading in the beaten track, not because there are no others, but because the traders are in general totally uninformed on those circumstances which lead to their discovery. The produce of the animal kingdom, in our commercial lists, is much more limited than that of the vegetable and the mineral. Yet how few of the valuable exotic drugs, dyes, and medicines do we know more of than their ordinary names! Some that, from being produced in small quantities, and in a limited district, bear a high price, may very possibly be abundant in adjacent countries, or might be transplanted and cultivated in other situations less remote and more convenient. It is the business of the merchant, if he aims at wealth, to discover new sources of commerce, of which he can reap the first fruits; but this will never be done, save by accident, without he is well informed respecting the productions, — whether natural or artificial — of other nations; in order that he may supply their wants, or import their produce. The truth is, that the profession of commerce embraces many branches of information, and even of science, which at first sight appear totally unconnected with it: and among these, natural history holds no inconsiderable station.

(83.) The pursuits of the agriculturist and of the planter bring them more immediately into contact with the productions of nature; and hence they are more especially interested in understanding their qualities. It is not only necessary to be well acquainted with the different vegetables grown or reared for economic purposes, but to understand the cause of the injuries they are subject to; and then to devise efficient remedies for those injuries. Here, also, is a wide field open for improvement and for discovery, and in which no information is so practically useful as that afforded by natural history. We are continually hearing of the failure of crops, and of attendant ruin. Now, in nine instances out of ten, these devastations have originated in the unusual abundance of some particular insect, which, from unknown causes, has appeared in great numbers. We contend not that the knowledge or the ingenuity of man could foresee such evils, or could totally counteract them; but experience has shown how much may be done, in many cases, both in the way of prevention and of cure. To do this effectually, however, recourse must be had to natural history. The cause of the injury being ascertained, the habits of the insect must be studied in all its different stages. What will prove more or less efficacious in one of these stages, will be totally useless, or will increase the evil, in another. Hence arises the necessity of ascertaining names and species; without which, no effectual steps can be taken.

(84.) If there required any striking fact to

show the intimate connection between agriculture and natural history, it would be found in the circumstances which attended the supposed appearance of the Hessian fly in this country; thus mentioned by Messrs. Kirby and Spence:—" In 1788, an alarm was excited in this country, by the probability of importing, in cargoes of wheat from North America, the insect known by the name of the Hessian fly, whose dreadful ravages will be adverted to hereafter. The privy council sat day after day, anxiously debating what measures should be adopted to ward off the danger of a calamity, more to be dreaded, as they well knew, than the plague or pestilence. Expresses were sent off in all directions to the officers of the customs at the different out-ports, respecting the examination of cargoes: despatches were written to the ambassadors in France, Austria, Prussia, and America, to gain that information, of the want of which they were now so sensible: and so important was the business deemed, that the minutes of council, and the documents collected from all quarters, fill upwards of two hundred octavo pages. Fortunately, at that time, England contained one illustrious naturalist, to whom the privy council had the wisdom to apply; and it was by Sir Joseph Banks's entomological knowledge, and through his suggestions, that they were at length enabled to form some kind of judgment on the subject. This judgment was, after all, however, very imperfect. As Sir Joseph had never seen the Hessian fly, nor was it described in any entomological system, he called for facts respecting its nature, propagation, and economy, which could be had only from America. These were obtained as speedily as possible, and consisted of numerous letters from individuals, essays from magazines, the reports of the British minister there, &c. One would have supposed, that, from these statements, many of them drawn up by farmers who had lost entire crops by the insect, which they professed to have examined in every stage, the requisite information might have been acquired. So far, however, was this from being the case, that many of the writers seemed ignorant whether the insect were a moth, a fly, or what they term a bug. And though, from the concurrent testimony of several persons, its being a two-winged fly seemed pretty accurately ascertained, no intelligible description was given, from which any naturalist could infer to what genus it belonged, or whether it was a known or an unknown species. With regard to the history of its propagation and economy, the statements were so various and contradictory, that, though he had such a mass of materials before him, Sir Joseph Banks was unable to reach any satisfactory conclusion." (*Introduction to Entomology*, vol. i. p. 51.) Nothing, as our authors justly observe, can more incontrovertibly demonstrate the importance of entomology, as a science, than this fact. Those observations, to which thousands of unscientific sufferers proved themselves incompetent, would have been readily made by one entomologist well versed in his science. He would at once have determined the order and genus of his insect; and in a twelvemonth, at furthest, he would have ascertained in what manner it made its attacks, and whether it were possible to be transmitted with grain into a foreign country. On data like these, he could have pointed out the best mode of eradicating the pest, or of preventing the extension of its ravages.

(85.) But if some acquaintance with natural history may be thus beneficial in the councils of the nation, still more essential is it to those who possess lands in our colonies, and who are desirous of making them profitable. We hear, for instance, of the worn-out state of the West India plantations; that the soil will no longer repay the expenses of cultivation; and that the introduction of sugar, rum, &c. from other countries, has brought ruin upon these. We know not how far these statements may be correct; but admitting them to be so, it may be fairly enquired, what efforts have been made to remedy them? why could not the aromatic spices of the East be equally well grown in the West Indies? and why has not the cultivation of the silkworm been undertaken in the Antilles, instead of leaving this enormous trade in the hands of the Asiatics? Why, again, are not efficient and scientific trials made for rearing the tea plant either in the West Indies or on the neighbouring continent? What obstacles exist against the cultivation of the vine and the olive, — two plants which we know personally will flourish in every possible variety of soil, — in these ill-fated islands; and thus establishing in them new and important sources of commerce and of wealth? In deciding these and similar questions, natural history becomes of the first importance. Since the only data upon which operations can be properly conducted, must be furnished by persons well versed in that science; accustomed to enquire into, and reflect upon, those kinds of facts, which none but a naturalist would ever think of. So strongly, indeed, were some of our West India proprietors impressed with the expediency of instituting enquiries of this nature, that a meeting was held, some few years ago, for the express purpose of discussing the subject. They even went so far as publicly to announce the name of the naturalist who was to be sent on this mission. For some reason, however, the scheme was abandoned; and although the reasons for its execution are even stronger now, than they were then, it cannot be expected, in the present agitated and unsettled state of these colonies, that it will be soon revived.

(86.) Let us now consider the case of another description of agriculturists—those who carry their capital and their industry abroad, for the purpose of settling in foreign countries. To them, an elementary knowledge of natural history is of much more consequence than to the English farmer, who frequently learns, from the experience of others, what is to be done in cases of emergency; or who can, at least, apply for such information to scientific advisers. But the agricultural emigrant has not these resources: he has, for the most part, to learn every thing himself: he has to study soils, and try experiments as to the crops best adapted to them. These crops will frequently be attacked and destroyed by a host of new enemies of the insect world, the species of which he has never before seen, and against which, in consequence, he knows not how to proceed. He is, in fact, thrown upon his own resources; and if he has not a sufficient knowledge of natural history to enable him to *reason* upon the facts before him, or to direct him how to proceed, he suffers the full extent of evils which might otherwise have been mitigated or prevented.

(87.) How continually are the nurserymen and gardeners of this country complaining of extensive damage done to their crops and their fruit-trees by different species of insects! Yet these very insects from being called by vulgar provincial names, are almost totally unknown to naturalists, who cannot, therefore, supply that information which is desired It is surely not too much to expect that a gardener should be able to tell the difference between a beetle and a fly; between an insect with four wings, and one without. Yet so little has this information been thought of among the generality of this profession, that not one in twenty has any knowledge on the subject. Country gentlemen complain of their fruit being devoured by birds, and orders are given for an indiscriminate destruction of birds-nests: the sparrows, more especially, are persecuted without mercy, as being the chief aggressors; while the Robin redbreast, conceived to be the most innocent inhabitant of the garden, is fostered and protected. Now, a little acquaintance with the natural history of these two birds would set their characters in opposite lights. The sparrows, more especially in country situations, very rarely frequent the garden; because, grain being their chief food, they search for it round the farmyard, the rick, and the stable: they resort to such situations accordingly. The Robins, on the other hand, are the great devourers of all the small fruits: they come from the nest just before the currants and gooseberries are ripe; and they immediately spread themselves over the adjacent gardens, which they do not quit so long as there is any thing to pillage. It may appear strange, as it certainly is, that no writer on our native birds should have been aware of these facts; but it is only a proof how little those persons,— who are, nevertheless, interested in knowing such things,—attend to the habits and economy of beings continually before their eyes. In like manner, we protect blackbirds for their song, that they may rob us of our wall and standard fruits with impunity.

(88.) It behoves every one to show humanity to animals, although we are authorised and justified in destroying such as are found, by experience, to injure our property. Under this latter head, however, we are committing so many mistakes, that, ere long, some of the most elegant and interesting of our native animals will probably be extirpated. Country gentlemen give orders to their gamekeepers to destroy all " vermin" on their preserves; and these menials, equally ignorant with their masters of what " vermin" are really injurious, commence an indiscriminate attack upon all animals. The jay, the woodpecker, and the squirrel,—three of the most elegant and innocent inhabitants of our woods,—are doomed to the same destruction as the stoat, the polecat, and the hawk. Nothing, in our native ornithology, can be more beautiful than the plumage of the jay: while its very wildness and discordance is in harmony with the loneliness of the tangled woods it loves to frequent. The sudden and sharp cry of the green woodpecker is of a similar character; and the sound of its bill " tapping the hollow beech tree" is interesting and poetical. The squirrel, again, is the gayest and the prettiest enlivener of our woodland scenery; and, in its amazing leaps, shows us an example —unrivalled among our native quadrupeds — of agility and gracefulness. Yet these peaceful denizens of our woods are destroyed and exterminated, from sheer ignorance of the most unquestionable facts in their history. The jay, indeed, is said to suck eggs; but this is never done except in a scarcity of insect food, which rarely, if ever, happens. The woodpecker lives entirely upon those insects which destroy trees, and is, therefore, one of the most efficient preservers of our plantations; while the squirrel feeds exclusively on fruits and nuts. To suppose that either of these are prejudicial to the eggs or the young of partridges and pheasants, would be just as reasonable as to believe that goatsuckers milked cows, or that hedgehogs devoured poultry. It is surely desirable that right notions should be had on such things, and that by an acquaintance with the most common facts of natural history, our few remaining native animals should be preserved from wanton and useless destruction. If natural history can teach us nothing more than humanity towards such inoffensive creatures, a little attention to it would not be misplaced.

(89.) We have now touched upon most of those subjects in which the study of nature may be brought to practical and beneficial purposes. Many of them may be thought trivial, and some remote; but there are others which involve questions concerning the prosperity of large communities, and the success of great commercial undertakings. No science, which can be applied to the solution of such questions, can be deemed inapplicable to the every-day purposes of life; or unconnected with the wealth of nations or of individuals. There is, in fact, scarcely any branch of human knowledge but what may be applied, immediately or remotely — in one shape or another, — to the common benefit of mankind; and among these, natural history, both in its moral and practical application, must ever hold a distinguished place.

(90.) To travellers in foreign countries, natural history is now become almost an essential qualification. In the infancy of the natural sciences, the productions of remote countries were either assimilated to our own, or magnified and distorted into the most marvellous wonders. The fabulous accounts of the natives were faithfully collected by the credulous traveller, and given to the world as facts attested by his own observation. Hence arose the absurdities recorded by Marco Polo, Ferdinand Mendez Pinto, and many of the earlier travellers, no less than the erroneous names assigned, in books of more modern date, to animals whose species never existed where they are asserted to live. But the advance of knowledge, and a more attentive consideration of animal geography, has shown us that these accounts can no longer be depended upon; and an ornithologist would no more expect to find the sparrow of Europe in the farmyards of the Cape Colony, or even in North America, than he would to discover a race of Indians in the mountains of Scotland. Now, as there are few countries out of Europe, where, if a traveller goes, he will not have to speak of its natural productions, it follows that his qualification for doing this will be measured by his proficiency in natural history. He may, indeed, omit the subject altogether; but it will be at the hazard of his book holding a very inferior station in the estimation of the public. We allude not, of course, to those entertaining but ephemeral narratives of travels, published under the appropriate titles of *Notes, Sketches, Short Residences,* &c. wherein amusement rather than instruction is aimed at. It is not to such sources that we are to look for solid information on the laws, the statistics, or the productions of a country; nor do we place them as standard books of reference on the same shelf of our library as Humboldt's New Spain, Burchell's Africa, or Ward's Mexico. Our observations are addressed to travellers of a higher class: yet even the sketching and noting tourists of the day, while they gallop over a certain number of leagues against time, would do well to know something of the animals which they pass, or the productions which they cannot stop to bring home as tests of their veracity. The world of animals is as replete with anecdotes as that of man; and although they may not be so generally amusing, they will often be found more instructive.

(91.) Natural history, indeed, now forms such an important feature in the best voyages and travels, that the subject is usually assigned to a professed naturalist, and is either made a separate division of the volume, or is published as a distinct work. By such arrangements, science gains the full advantage of the discoveries made, for they are generally given to the world by those most competent to the task. But this, so far from lessening the necessity of natural history forming one of the accomplishments of the traveller, rather increases that necessity. The science, as before observed, can only be prosecuted with full advantage by two classes of students; pursuing, indeed, the same end, but attaining it by different means. The practical naturalist studies in the fields; he collects specimens, he observes instincts, and he records facts. His scientific brother compares these acquisitions with those already existing; he studies organisation, and he consults books. Both these modes of investigating nature are essential to the true knowledge of her works; but they can seldom be prosecuted by the same individual. In this, as in almost every branch of science or of art, the advantage of the division of labour is manifest. To the traveller, therefore, belongs the first set of these duties. If he has no intention of publishing, himself, a detailed and digested account of his discoveries, but is desirous that others more experienced should do so, he has yet to understand the practical part of his subject. The art of preserving specimens, and some little knowledge of the science, must be first acquired, before he can judge what species to reject and what to preserve. The economy of animals, again, can only be learned by observation; and this implies a habit of quick-sighted attention, and a knowledge of such points as should be more especially attended to. Facts, so trivial in themselves that ordinary observers would pronounce them insignificant, are often, in the eye of the naturalist, of the highest importance; not, indeed, in their isolated character, but as leading to or corroborating some of the great truths of the natural system. As an instance of this, the manner in which the chrysalis of a butterfly is suspended, whether with its head upwards or downwards, would appear, to all but an entomologist, too trivial for record. Yet this simple variation of position determines at once to which of the primary types of the diurnal *Lepidoptera* the insect in question belongs. A thousand similar instances might be adduced, were they necessary, to enforce the most critical precision in recording observed facts.

(92.) The advantages of natural history, as a philosophic study, need not be dwelt upon, after what has been so ably said in reference to the physical sciences in general*, of which this forms but a part. We have already shown, that there are departments which may be cultivated without a profound acquaintance with those physical laws hereafter to be explained. But were we, in this place, to enumerate all those qualifications which constitute a philosophic naturalist, — the varied acquirements he should possess — the materials he should collect — the years that he must study — the countries he

---

* See Sir J. F. Herschel's Preliminary Discourse.

should visit,—we fear the reader might interrupt us with the exclamation of Rasselas to Imlac*, while the latter was proceeding to aggrandise the profession of a poet. "Enough! thou hast convinced me that no human being can ever be"— a naturalist.

---

* Johnson's Rasselas, chap. xi.

PART III.

OF THE PRINCIPLES ON WHICH NATURAL HISTORY RELIES FOR ITS SUCCESSFUL PROSECUTION, AND THE CONSIDERATIONS BY WHICH THE NATURAL SYSTEM MAY BE DEVELOPED.

CHAPTER I.

ON THE DISMISSAL OF PREJUDICE.

(93.) It has been truly and forcibly urged*, that the dismissal of prejudice is absolutely essential to the prosecution of science: and we may add, that if there be any branch of physical knowledge which more especially calls for this dismissal; or whose progress, more than that of any other, has been impeded by prejudice; it is that of natural history. We allude more especially to prejudices of opinion; since those of sense, however they may arise in other sciences, are subordinate to this. Natural history is a science of facts and of inferences. The former regard structure and economy; and as these, under favourable circumstances, can be investigated

---

* Sir J. F. Herschel's Discourse, p. 80.

by every one, few prejudices of sense can arise respecting them. But when we proceed further, and attempt, from these facts, to draw inferences, the case is different. No principles having been yet established, by which the facts we know from experience can be generalised in such a way as to establish their mutual relation and dependence. Every naturalist therefore thinks he is at liberty to draw his own inferences, and to apply them to the systematic arrangement of the objects by which they are furnished. One, for instance, arguing from the flight of the bat, looks on it as that animal which constitutes the true passage from quadrupeds to birds. Another, looking to its general aspect, is disposed to place it among the mice, fortified by the general name given by the French to the whole tribe of *chauve souris*. A third, chiefly influenced by the peculiarity of its teeth, arranges it in the same group as the monkeys: and each, acting upon his respective inferences, fashions his system accordingly. Now, as to the facts connected with the individual structure and the economy of the bat, all these naturalists would agree; because such facts can be verified by their personal observation, and there would be no room for prejudice. But here unanimity ceases. They proceed to inferences; and each, laying a peculiar stress upon some one fact more than upon others, makes it a principle of his own arrangement. This is the true cause of the number and the mutability of zoological systems. In respect to the bat, it is very clear, that if there is an order or progression in nature, — which no one ever thinks of doubting, — this quadruped can hold but *one* station in the scale of being; and it therefore follows that only *one* of the opinions just glanced at can be true. Now, if we are unacquainted with any general laws of animal variation, by which the soundness of conflicting inferences can be tested, how are we to decide in the case before us? It is clearly impossible: each opinion is supported by reasons, and each party appeals to and acknowledges the very same facts. In the infancy of science, such questions were generally decided by the authority and the influence of a name. As knowledge increased, such arbitrary authorities also multiplied; but their influences proportionally declined. Each, however, still continues to have its little circle of disciples, who, from having studied under, and imbibed the system and opinions of, their master, tenaciously adhere to what they have been taught to consider as truth.

(94.) Here, then, lies that species of prejudice against which we would more especially caution the student; and which, if he will not conquer it, will incapacitate him, both from rising to the present level of science, and from extending its boundaries. He should ever bear in remembrance, that facts, authenticated by the experience of others, or falling under his own cognisance, are immutable, because nature is ever the same; but that the inferences from them may be so numerous, and so contradictory, that, until we are acquainted with some general laws whereby universal agreements can be established, one inference, in point of fact, is just as good as another. To illustrate our meaning more plainly, let us look to four of the greatest authorities on the classification of the *Mammalia*, namely, Linnæus, Cuvier, Illiger, and Hamilton Smith. Each of these studied from the same models,—models which are now the same as when they were first created; and each and all agree in the results of their respective examinations; that is, in the facts belonging to the structure of these animals. So far, therefore, we need not question their authority. But when they began to reason upon these facts, each drew separate inferences, and consequently produced different systems or methods of classification. These systems, however, make no reference to other parts of creation. They treat of the class before them, as if it was the only one in nature, and as if the principles by which it was to be arranged had no connection with those which governed other classes. We find, in short, no allusion to mutual resemblances out of this division of animals; so that an ordinary reader would suppose that nature had one system for quadrupeds, another for birds, a third for fish, and a fourth for insects. Did he turn to the best classifications of each of these orders now in use, he would be still further confirmed in this opinion, by seeing that they were all treated of in the same isolated and disconnected manner. The ornithological systems of the greatest naturalists in this department differ from each other fully as much as those relating to quadrupeds, and are calculated to produce the same impressions. Seeing, therefore, that inferences may be innumerably various and discordant; by what rule, as the science now stands, are we to be guided in choosing truth? It is evident, that if there be but one system in nature, there can be but one natural mode of classifying her productions; that is, in the true series or chain of being. We venture these remarks, not in disparagement of the great names who have gone before us; and to whom, on so many other points, science is highly indebted; but that the naturalist may clearly see the shallow basis upon which his prejudices of opinion rest, when they have been formed in favour of isolated systems and arbitrary methods. Upon a subject, however, of so much importance to the successful prosecution of science, we may offer some further considerations.

(95.) It is a fact which the progress of human knowledge has demonstrated, and which is continually receiving new and corroborative proofs, that the more we understand of the primary laws of nature, the more simple, universal, and harmonious do we find them. To suppose, therefore, that a theory of arrangement can be natural, which pretends not to explain and to illustrate any one general law of nature, is, in fact, either virtually to deny that any such exist, or that, however other sciences may be governed by general laws, that of natural history is exempt from them. To adduce arguments against either of these propositions would be a waste of words: their futility being admitted, how strongly will such considerations shake our faith, and destroy our prejudices, in favour of such systems of nature as are not founded on general laws. It may, perhaps, be admitted, that analogical reasoning authorises the supposition, that natural history, in this respect, differs not from other physical sciences; but it may be contended, these new theories, which have been recently promulgated, and which their discoverers designate as *natural*, have, as yet, been but imperfectly explained; that, at present, they are crude and ill-defined, and consequently, that they are too imperfectly developed and too partially verified, to merit general confidence. Yet to whom is this latter fault to be attributed, but to those who urge it? If the advocates of arbitrary classification contend that it will be time enough to dismiss our present systems when these new theories have been extensively proved in every department of zoology, and yet refuse, themselves, to join in the Herculean task, and to try how far these new views can be verified in unarranged groups, they contribute to augment that evil of which they complain: and if they thus determine to evade this labour, we can scarcely hope that any thing effectual will be accomplished in the present century. How much better would it be for science, if, instead of urging such querulous complaints, these advocates for what is old would overcome prejudice of opinion, and resolve to try every theory that professes to develope general laws by the surest of all tests — their universality. At all events, even if we allow the full force of their objection, the only just inference to be drawn is, that our prejudices in favour of arbitrary systems should be shaken, if not overcome. He who considers that natural history is to be studied by rules different from those by which all other physical sciences are prosecuted, is totally unfit to meddle with it.

(96.) Prejudices of opinion, also, in regard to natural history, are to be combated by another consideration. No one, who believes that the creation is the work of Omnipotence, can for a moment suppose that it was called into being without some great plan, or method. It therefore follows, that, unless our exposition of such parts of this plan as we believe we have discovered, is found equally applicable to *all* groups in nature, there is inductive evidence to believe that our theory is fundamentally wrong. It is contrary, as we before observed, to the sense of the word *method*, that quadrupeds should have been created on one system, birds upon another, and insects on a third. The harmonies of the natural world are every where conspicuous; and how can we suppose that the most perfect works of the Creator, save and except man, have been framed without any regard to unity of plan, and harmony of purpose? The supposition is monstrous, and not to be admitted for a moment. This alone should be sufficient to shake our prejudices in favour of all such systems or theories as are made applicable to one division of nature, but not to the others.

(97.) Having now, as we hope, sufficiently warned the student against prejudices of opinion, by pointing out to him those rules by which the value of all systems and theories regarding his favourite science are to be judged, we shall advert to those few prejudices of sense which belong to this science; and which, however trivial they may appear in themselves, may be productive of essential injury to science when used as arguments.

(98.) Prejudices of sense, in natural history, are chiefly confined to opinions derived from witnessing animals under conditions of existence not habitual to the species; and which, from being casual and incidental, may be termed unnatural — that is, contrary to their usual natures. First, as regards the *habits* of animals; upon which, as will be hereafter seen, their station in the scale of nature so much depends; I shall adduce a striking instance, which might have given rise to a prejudice of this sort, in a case witnessed by myself this year. A particular tree on the lawn, immediately opposite my library window, is the usual station of two grey flycatchers, who have frequented it annually for the last five years. Those who are acquainted with the manners of this bird, know that it habitually lives seated upon trees, where it remains stationary, darting occasionally upon passing insects, and returning to the same twig, without perching on the ground. This peculiarity of habit is confirmed by its organisation: for, on looking to the feet of the bird, we see they are of such a construction as to incapacitate it from habitually walking, or even hopping, upon the ground. Nevertheless, I observed this year, for the first time, one of these birds in such a situation: it was upon the grass but once; and then, apparently, to secure an insect which it had wounded, but not captured, in its first assault. Now, had this fact been witnessed by an observer, not acquainted with the ordinary habits of the species, and ignorant of the influence which structure exercises upon *habits*, he would at once affirm, from his own personal experience, that the grey fly-catcher, and, consequently, all the species of the same genus, were *in the habit* of frequenting the ground. Here, then, is a case where we must even distrust the evidence of our senses, if we are tempted to apply that evidence too hastily to the generalisation of facts. The habits of almost every animal, even in a state of nature, if attentively watched, might probably furnish instances, equally strong, of occasional aberrations from that economy which, to them, is natural and habitual. Such incidental facts must be viewed under the same light as we regard monstrosities, or *Lusus Naturæ;* and the only legitimate inference we can draw from them is the futility of all *absolute* characters.

(99.) But if prejudices may be imbibed from viewing animals in a state of nature, still more may they be generated by looking to animals in confinement, and drawing inferences from the habits or instinct they then exhibit. A curious instance of this has just been published. With a view to ascertain the natural food of the hedgehog, an individual was confined in company with a snake. As might naturally have been expected, the hedgehog, when pressed nearly to starvation, attacked and devoured the latter: the fact was undeniable, and the inference deduced was, that nature intended this quadruped to prevent our being overrun with serpents. Against this conclusion it has been urged, that the one is only abroad during the day, while the other feeds only by night; so that by no *ordinary* chance would they ever meet. (*N. H. Mag.*) Did the hedgehog, like the mole, habitually burrow in search of its prey, we might then, indeed, conjecture that it dug out serpents from their holes during night: but this supposition, again, is not borne out by anatomical structure; and we are at last compelled to believe that the fact before us, although true, is contrary to the usual course of nature.

(100.) Prejudices of sense show themselves in various ways. A recent author, who writes upon shells, finding a new genus of barnacle, occupying the deserted holes of some perforating bivalve*, concludes that the excavations have been made by the barnacle. and therefore names it *Lithotrya!* There is nothing surprising or reprehensible in this, for it is an error which all who trust alone to their eyesight would most assuredly have fallen into. On looking further into the matter, however, and ascertaining the general structure of this group of animals, so admirably illustrated by Poli†, we see the physical impossibility of their possessing this perforating power; and it is known, moreover, that barnacles, instead of being shell-fish, are articulated *Annulosæ*, and belong to the class of Insects. ‡ The origin of the genus *Gionea* has recently been quoted by one of the most eminent mathematicians of the age §, as an example of fraudulent *hoaxing;* although, I confess, it appears to me more allied to the subject now touched upon. M. Gioeni finds, upon the coasts of Sicily, the hard internal parts of a shell-fish (*Bulla lignarea* L.); and this object, more resembling a bivalve than any thing else, he mistakes for the shell itself, and publishes it as such. The cele-

---

* Cuvier, Règ. Animal, 2d ed. vol. iii. p. 177.
† Testaceæ Utrinque Siciliæ, vol. i. pl. 4, 5, 6.
‡ See also Thompson's Zool. Researches, No. 3.
§ Babbage on the Decline of Science, p. 175.

brated Bruguire, deceived in like manner by using only his eyesight, adopts the same idea, and places* this supposed new genus close to that of *Pholas*, to some species of which it certainly bears no small resemblance.

(101.) Perhaps, the most inveterate of all these sorts of prejudice is that which induces people to believe that frogs and toads can live for centuries in blocks of marble, impervious to air and of course to food. We are so repeatedly assured of this fact by writers in newspapers and periodicals, wherein all the circumstances, with names and dates, are given, that nothing but an actual series of experiments could demonstrate the truth or falsehood of such an alleged departure from the known laws of nature. Such experiments have accordingly been made, and the results have been just what might have been expected by any one accustomed to inductive and analogical reasoning. Yet, had not the trials here alluded to been made, it might have occurred to us as a singular fact, that out of so many recorded instances of toads being found in stones, no specimen of the broken *nidus*, and of the antediluvian reptile alleged to have been within, has never been submitted to the inspection of the scientific. Nothing would be more easy than to collect the fragments of the one, and preserve the other in a bottle of spirits. We hope, therefore, that the first of our readers, who is within a short distance of such a discovery, will take this hint, and, by sending us the toad and the stone, silence

---

* Ency. Méth. pl. 170.

for ever our present obstinate incredulity on such wonders.

(102.) We have dwelt the longer upon the necessity of conquering prejudices in this science, because it occupies at this period a very peculiar station in the circle of human knowledge. All those leading naturalists who enjoy the highest rank in public estimation, agree in confessing that there must be general laws of classification; yet scarcely one has hitherto attempted to define what they are, or how they would act,—in other words, what results of harmonious combinations would follow their application. Every one agrees that there must be a natural system; yet no one has yet presumed to say what are the primary laws of that system. When, therefore, we venture to do this, — when we call to mind the weight of opinion that will be brought against us, the great names that have gone before us, and those which still live high in the estimation of nations and of their rulers,—we feel all the difficulties of our task, and that we have more than ordinary prejudices to encounter. Let him, therefore, who, from the force of habit, of early initiation into the reigning systems, or from being the author of one himself, finds himself incapable of patiently weighing arguments intended to overthrow his favourite theories,—let him, we repeat, close our volume: for in it he will find but little to interest him. We address ourselves to those who have been instructed to form more enlarged conceptions of the physical sciences; who view natural history but as a part, and considers that that part must be studied upon the same principles as any other. Great revolutions in science are scarcely ever effected but after their authors, and the generation to which they belonged, have ceased to breathe. Yet there is nothing unnatural or unaccountable in the slowness of this removal of error. After all, the authority of names, in questions of pure science, is not what it used to be. It may, indeed, for a time, operate against the diffusion of truth; but truth, once discovered, stands in no need of such aid. During the age when the zoological world bowed with unhesitating submission to the opinions of the great naturalist of Sweden, it was affirmed by him, and believed by the world, that corals were plants, and that swallows passed the winter under the ice.* Such prejudices are now only to be laughed at: but we may fairly enquire whether many of the opinions we now hold, will not equally excite a smile from our successors.

---

* The celebrated Peter Collinson thus writes to Linnæus, when opposing this latter prejudice: — " Your reputation is so high in the opinion of the learned and curious of this age (1762), that what you assert is taken and allowed to be a real fact; for when I have been reasoning on the improbability of swallows living under water, it has been replied, *Dr. Linnæus says so, and will you dispute his authority."— Lin. Cor.* vol. i. p. 54, 55.

CHAP. II.

ON THE PRINCIPLES ON WHICH NATURAL HISTORY, AS A BRANCH OF PHYSICAL SCIENCE, IS TO BE STUDIED.

(103.) THERE are two modes by which our knowledge of natural history can be successfully prosecuted. The first of these is to commence with investigating the forms and properties of *species;* combining them, according to their degrees of similarity, into groups or assemblages of different magnitudes; and then attempting to discover what general inferences can be drawn from such combinations, or, in other words, what are the principles by which their variations are regulated. This is the analytical method, by which we commence, as with an alphabet; and from letters determine words; from words proceeding to sentences; and, combining these, again, to chapters. By the second mode, we proceed quite differently. We begin by taking for granted the correctness of certain given principles, and apply them to the investigation and arrangement of some particular group. This is the synthetic mode. By the first, we commence as if all general laws were yet to be discovered; by the latter, as if they were already known, and only required a more particular or extended application.

(104.) As all true knowledge of the combinations of nature must originate in *analysis*, we shall first intimate how this can be most successfully prosecuted.

(105.) If we reflect for a moment on the sort of information which it is the province of natural history to teach, we shall find that all the knowledge of an organised being which it is possible to acquire, is comprised under one or other of the following heads: —
 1. Its structure and composition.
 2. Its properties.
 3. Its relations to other beings.

(106.) Hence it naturally follows that a knowledge of species is the true basis upon which the science reposes for its successful prosecution. We cannot combine objects, with due regard to their fitness, until we understand their structure and properties; any more than we can acquire a language, before we become acquainted with its alphabet.

(107.) A knowledge of structure, and of properties, is to natural history, what experience is to other branches of physical science. In either case, we look not to causes, for they are beyond our comprehension; but we look to objects or to facts, which every body, under favourable circumstances, can verify; and which, in consequence, become immutable truths. Upon this basis, therefore, we must commence the study of nature: and proceeding step by step, — measuring back our ground when we begin to doubt, yet gaining confidence from every corroborating evidence, — we advance from the foundation to the portico.

(108.) 1. Let us first enquire what are the considerations which enter into the structure and composition of a species. These considerations, again, may be classed under three divisions, viz. external organisation, internal anatomy, and chemical composition. The first of these belongs more especially to the zoologist, the second to the anatomist, and the third to the chemist: all are fit objects of enquiry; but as all are not equally essential to our present purpose, we shall confine our observations chiefly to the first.

(109.) For the sake of simplification, the word *form* or *structure* may commonly be used as synonymous with external organisation. If a person, unacquainted with natural history, was put into an immense store-room, filled with all sorts of plants, animals, and minerals confusedly mixed together, and then desired to sort and separate them, he would, even were he a clown, begin to place the plants in one heap, the animals in another, and the minerals in a third. If, after this was done, he was again directed to make a more particular assortment of the animals only, he would assuredly separate the quadrupeds from the birds; and these, again, from the fishes and the serpents. No one will deny that this would be the natural process: and we may therefore infer, that external form is the chief and primary mode by which nature herself teaches us to know her productions: and that we need only descend to an examination of internal structure, when this resource fails, and we are obliged to enter upon minute and delicate investigations. But as many have laid an undue stress upon the importance of *internal* over *external* organisation, and thereby, as we conceive, embarrassed the path of the student with unnecessary difficulties, it may be as well to explain, before proceeding further, in what respect these may be said to differ.

(110.) By form, or external structure, we comprehend not only the different external parts of the body, and of the members thereunto attached; but all such organs as have one of their surfaces, at some time or other, protruded and exposed to the eye, and which may be observed without the necessity of dissection. Hence it follows, that the jaws, the teeth, and the mouth of quadrupeds; the bill and tongue of birds; the *instrumenta cibaria*, or parts of the mouth, in insects; the external coverings of the bodies of tortoises (*Chelonia*) and shell-fish (*Testacea*); the retractile *tentaculæ*, where they exist, of caterpillars and snails; and the proboscis of inferior animals;— all these are parts of their external anatomy, or, as we shall hereafter say, of their *form*. All other, — that is to say, such as are enveloped and concealed beneath the cuticle, or that substance which acts as the external protection of the animal, — relate to its *internal* construction or anatomy. We contend not for the critical accuracy of these definitions, but for their general truth and convenience. It may be urged, indeed, with some show of reason, that nearly all teeth are internal, and that the organs of the mouth in insects cannot be studied without dissection; but it must likewise be remembered, that neither one nor the other are enveloped in other substances, and that their outward surface is exposed to the action of atmospheric air.

(111.) To estimate aright the respective value of these two modes by which animals may be distinguished, becomes the first duty of the student;

and he is, therefore, led to enquire how far they have been used by others? and which of them is best calculated to aid his studies? Where two modes of investigation conduct us to the same results, it cannot for a moment be questioned that the preference should be given to that which is most simple; for this preference not only abridges individual labour, but tends to render science inviting to others. To define disagreements, and to point out similitudes, is the chief business of the naturalist; and if he can accomplish this, the world will be satisfied and convinced, in proportion as the means he has employed, or the arguments he has used, can be verified and understood by others. Suppose, for instance, that the physiologist, who wished to inform us on the different varieties of man, directed our attention — not to the external peculiarities of their features, which every one can see and comprehend — but to the different modifications of their internal anatomy, which could only be understood by one reader in a hundred: our question would immediately be, Why have recourse to these complex characters, when others, apparent to the most illiterate observer, lie before us? A European can be as accurately distinguished from an Ethiopian by his external form, as by the most refined specification of any anatomical differences that may exist between them; while, if there are no such internal differences, we become persuaded that the variations of nature can be best understood, and can be more accurately defined, by the more simple and natural process of studying her external distinctions.

(112.) A system in which it is professed to arrange all animals *according to their organisation*, certainly carries with it an imposing aspect of authority, originating not so much in the high reputation of the author as a comparative anatomist, but from the supposition conveyed by the title, that it is based upon the *internal* organisation of the animal kingdom; and that all the divisions are formed with a primary regard to such considerations. But what is the real fact? Where one group of animals has been dissected, and their internal structure explained, there are twenty which are defined only from their external appearance; so that, with the exception of occasional notes introduced as subordinate characters, we find that by far the largest proportion of the details of this system is founded alone upon external form; and that these external characters are nearly as much insisted upon in the *Règne Animal* as in the *Systema Naturæ*. Further than this, indeed, the comparison between these admirable works cannot be carried, except that each commenced a new era in that science which they have so signally benefited.

(113.) To external form, then, we must chiefly resort, if we wish to make the productions of nature intelligible to the generality of mankind. We have seen that, in the case of man, nature has chosen external peculiarities to mark the distinction of different races; and it is notorious how universally she has employed the same means to distinguish those millions of individuals who people the earth; so that, by innumerable modifications of the same set of features, we are able to recognise a countenance with which we are familiar in a crowd of others, born of the same race, and of the same country. If we look to the animal world, the same results are apparent, more especially in all the leading systems on the vertebrated classes. In ornithology, particularly, there is not a single division which has been mainly founded upon internal structure: so that, in this class, even the *Règne Animal* follows precisely the plan of the *Systema Naturæ*. Hence we may conclude that external characters are almost always preferable to those founded solely upon internal structure; and that this conclusion is tacitly admitted by those who, in theory, maintain a contrary opinion.

(114.) Nevertheless, it would be absurd to suppose that the internal construction of an animal is not deserving of great attention. This study, in fact, constitutes, of itself, a distinct branch of physical science; useful, indeed, to the zoologist, as the means of assisting and guiding his studies, but by no means so essential as is generally supposed. Wherever external peculiarities are sufficient to supply us with clear definitions, we require no other. It must nevertheless be remembered, that when we descend to the lower groups of animal life, where the forms become proportionally simple, we must then have frequent recourse to dissection; not so much, indeed, for the purpose of characterising such forms, as for that of ascertaining to which of the grand divisions of the animal kingdom they truly belong. We allude more particularly to those soft molluscous-like animals, confusedly put into the Linnæan class of *Worms*. These stand so low in the scale of creation, that many of them have the aspect of plants and flowers. It is in the investigation of these beings, that some acquaintance with comparative anatomy is essential. Having now sufficiently discussed the relative value apparently belonging to different parts of the *structure* of an animal, little need be said on its composition. This, in fact, is the province of the chemist, whose business it is to analyse, not the form, but the elements of which that form is composed. Such considerations, no less than those belonging to internal structure, are essential to the full and complete knowledge of an organised being; but, whenever such a being can be defined with sufficient accuracy by more simple means, a redundancy of knowledge and a complication of characters are clearly to be avoided.

(115.) 2. Let us now pass to the second head of our subject: viz., the *properties* of an animal. It is evident, to an attentive observer, that the innumerable beings composing the animal creation are destined to perform different offices therein; and that they are not only endowed with *forms* adapted to such offices, but with *instincts* for carrying them into effect. Our attention is naturally directed, in the first instance, to their *forms*, because they may be understood and recognised long before we become acquainted with the designs for which such forms were created. The properties, therefore, of an animal consist, first, in its habits or instincts; and secondly, in the mode in which the qualities contribute to the general economy of nature.

(116.) The economy of nature, — that is to say, the harmonious adjustment of all created things,—is preserved by the efforts of all, instinctively directed, by an infinity of ways, to one end. If, to effect some purpose unknown to us, this nice adjustment (but in a single instance) is for a time suspended, we see disorder, devastation, and ruin inevitably follow. Such instances are not rare, for they are continually brought before us; and they may be looked upon as examples of what would follow, if there was no supreme Superintendence over creation. No insect is better known that the cockchafer, so common during summer. It feeds upon foliage; yet, in ordinary years, its numbers being regulated and kept within due limits, it is in no way injurious. Instances, however, have occurred, where these restrictions upon its increase appear to have been suspended, and the consequences were fearful. In the year 1688, immense hosts of this beetle suddenly appeared in Ireland: all vegetation was covered and destroyed by them; so that, but for their timely removal, famine would have fallen upon the land, and a pestilence have arisen from their dead bodies. We cannot doubt but that this and similar instances form part of the economy of nature, and are connected with causes and effects far beyond our penetration; but we must still consider them as deviations from the ordinary course of things, and from those rules by which we are accustomed to judge of the harmonious regulation of the universe.

(117.) Our first object, however, after becoming acquainted with the *form* of an animal, is to ascertain its habits and economy; without which it will be impossible to speculate upon its *uses*, or to understand in what way it promotes the harmony we have just alluded to. Now, this knowledge is to be acquired in two ways; either by actual observation, or by inductive reasoning. The first, of course, is the most simple and the most complete, and lies within the reach, under favourable circumstances, of every observer. The latter, on the other hand, is more confined, and can only be arrived at by a long course of study. The former is a mere exercise of vision; but upon the latter we have generally to reason analogically. We then find that certain modifications of form indicate certain habits; and that this reciprocity is so universal, that we are enabled to decide whether a bird, whose skin only we have seen, lives in general upon the ground, or among trees; whether it eats insects, or seeds, or both; or whether a beetle feeds upon green or upon decomposed vegetables.

(118.) To illustrate the importance of that minutely accurate observation which is necessary in ascertaining the habits and economy of animals, and at the same time to exemplify the diversity of ways by which nature effects the same object, let us imagine a noble forest tree, in whose luxuriant foliage the birds of the air find shelter, and whose leaves supply food to hosts of insects. In this respect, the tree may be considered a world in itself, filled with different tribes of inhabitants, differing not only in their aspect, but even in the stations or countries they inhabit, and assimilating as little together as the inhabitants of Tartary do with those of England. First, let us look to those insects, which, being destined to live upon vegetable food, are instinctively directed to seek it here: some, as caterpillars, feed upon the leaves; others upon the flowers, or on the fruit; a few will eat nothing but the bark; while many derive their nourishment only from the internal substance of the trunk. Every part of the tree is thus seen to supply food " in due season" to all these diversified tribes. If we examine further, new modifications of habit are discovered. Those insects, for instance, which feed upon leaves, do not all feed in the same manner, or upon the same parts: a few devour only the bud; others spin the terminal leaves together, forming them into a sort of hut, under cover of which they regale, at leisure, upon the tenderest parts; some, apparently even more cautious, construct little compact cases, which cover their body, and make them appear like bits of stick, or the ends of broken twigs; some eat the outside of the leaf only, while others,—like the caterpillars of New Holland, mentioned by Lewin, —bore themselves holes in the stem, into which they carry a few leaves—sally out during the night for a fresh supply, and feed upon them at their leisure during the day. It seems, in fact, impossible to conceive greater modifications than are actually met with, even among insects which feed only upon leaves; while other variations are equally numerous in such tribes as live upon other portions of the tree. Hence it is apparent, that in accurately determining the habits and economy of insects, no less than of animals generally, the greatest nicety of observation is absolutely essential.

(119.) If we look to the feathered creation, we shall find an equally remarkable diversity of habits, even among those tribes whose food consists entirely of insects. To illustrate this, even in a very confined compass, let us still fancy the tree we have just spoken of, bearing in itself a living world of insects, yet flourishing in beauty and luxuriance. We might imagine that the innumerable artifices by which these little creatures are taught to guard themselves, would effectually protect them from their enemies; and that, so secured, they would go on to " increase and multiply" with that rapidity which naturally results from security. But what would then be the inevitable consequence? Certainly the death of the tree, by which the whole are fed! for if these devourers of leaves, of flowers, of fruits, of bark, and of sap, were doubled or trebled, which they very soon would be, both tree and insects would perish together. Now, that this general destruction should not happen, but that the lives of *another* class of animals should be supported by the superabundance of the insects, birds are called into being, and are appointed to fulfil their repective parts in the wonderful economy of nature. Let us, then, look to those tribes who would frequent this same tree for the purpose of seeking food; and who would thus, by so doing, prevent the catastrophe we have just supposed. The woodpeckers (*Picianæ* Sw.) begin by ascending the main trunk; they traverse it in a spiral direction, and diligently examine the bark as they ascend; wherever they discern the least external indication of that decay produced by the perforating insects (generally the grubs of beetles), they commence a vigorous attack: with repeated strokes of their powerful wedge-shaped bill, they soon break away the shelter of the internal destroyer, who is either dragged from his hole at once, or speared by the barbed tongue of his powerful enemy. Next come the creepers and the nuthatches: they have nothing to do with these tribes of insects just mentioned, which are the peculiar game of the woodpecker: *their* food is confined to the more exposed inhabitants of the bark; the crevices of which they examine with the same assiduity, and traverse in the same tortuous course, as do the woodpeckers: the one taking what the other leaves. It is remarkable, that in temperate regions, like Europe, few insects are found on the *horizontal* branches of trees; and this seems the true reason why we have no scansorial birds which frequent such situations: but in tropical countries the case is different; and we there find the whole family of cuckows exploring such branches, and such *only*. Finally, the extreme ramifications, never visited by any of the foregoing birds, are assigned, — in this country at least, — to the different species of titmice, whose diminutive size and facility of clinging are so well suited for such situations. In this manner are the insect inhabitants of the trunk, the bark, and the branches, kept within due limits; while those which frequent the leaves become the prey of other birds. The caterpillar-catchers of Africa, India, and New Holland, as the name implies, feed only upon the larger sized larvæ; while, in this country, the whole family of warblers make continued havoc among all those lesser insects which live among foliage. Wherever, from climate or local situation, insects are most abundant, there also are the agents for subduing them proportionably increased. Thus, in America, the warblers are particularly numerous, and not only feed upon creeping insects, but also upon those winged tribes which frequent the foliage for shelter. The more we see of the economy of animals, the more do we find stratagem opposed to stratagem: so that modes of defence or of self-preservation, which even the reason of man would suppose perfectly effectual, are still found to be unavailing, in all cases, against the address of those enemies whose attacks are to be dreaded. In the foregoing remarks, we have been insensibly led to illustrate more than one of the positions before touched upon. The student, however, will thus perceive, to the full extent, the indispensable necessity of observing and recording every fact, even the most apparently trivial, connected with the habits of animals.

(120.) But, however attentively we may study the manners of living animals, there must ever remain a large proportion whose economy has never been recorded, and of which we can consequently know nothing from actual observation. Here, then, we must have recourse to analogical reasoning. It is found that certain habits are always indicated by a correspondence of structure. Among birds, for instance, we observe that all those which live habitually upon the ground, like the partridge, the turkey, and the domestic fowl, have strong, elevated legs; while in birds which rest only upon trees, like the swallow, these members are short and weak. We know, from experience, the universality of these facts; and we thence conclude that all birds so constructed, have corresponding habits, although we have never seen them alive, nor have any testimony to the absolute fact from others. It is obvious, again, that birds or insects having long and very pointed wings, are endowed with great powers of flight; for we see this structure is universal among the swallows, the humming birds, and the dragon flies. We can be in no danger, therefore, of mistake, in deciding on the slowness or quickness of flight in a bird, although we may never have seen it alive. The form, therefore, — or, as it is sometimes called, the *conformation* of an animal, — will generally reveal, to the experienced naturalist, the leading points of its natural economy. But these deductions can only be arrived at when the student has made considerable proficiency in the science; and has, by a diligent comparison of the structure of species with reference to their natural economy, duly qualified himself for forming opinions which have not yet been confirmed by his own observation. Leaving this subject, therefore, as more suited to another part of this discourse, let us enquire into the second division of our subject; namely, the properties of animals in regard to their influence or uses in the economy of nature.

(121.) So far as we have hitherto proceeded, the knowledge to be acquired results from simple experience; that is, it regards isolated facts, upon which there can be no dispute. The structure of an animal, and its habits or manners, are independent of all theory; and, when fully ascertained, are so many truths which may recorded in our chronicles, and appealed to by all parties as matters of indisputable authority. But, when we come to enquiries touching the purposes for which an animal has been created, and what influence it possesses in the economy of the universe, we pass the boundaries of simple fact, and are compelled, in most cases, to rely upon theory. True it is, as in the instances just given, we can be at no loss to discover the more general uses of animals: for example, we know that some supply food to others, or hasten the decomposition of decayed matter; that some promote the fecundation of plants, or check the exuberance of vegetation. These may still be admitted in the list of unquestionable truths, because they are manifest to ordinary observers. But when we enquire into more minute particulars, and speculate on the reasons why the flamingo, for instance, has such disproportionately long legs; what particular purposes of nature are fulfilled by the ostrich; or what are the particular uses for which such an apparently anomalous animal as the ornithorynchus, — half quadruped, half bird, — was created; when, in short, we attempt to discover the *uses* of such animals of which direct evidence cannot be produced; we enter upon a boundless region of speculation and theory, — a region which the student should avoid, and where the more experienced naturalist will do well to proceed with caution. In the mean time, such considerations should not deter us from accumulating facts connected with animal economy, or from recording such inferences as may be plausibly drawn from them; leaving the validity of these inferences to be confirmed or disproved by longer experience.

(122.) There is a third consideration regarding

the properties of animals, which should here claim our attention; for although it is disconnected with abstract science, and is not essential to the discovery of general principles, it is yet highly interesting to the bulk of mankind, because it concerns their individual interests. We allude to such properties of animals as are hurtful or beneficial to man; which are to be counteracted from being pernicious, or turned to our advantage from their usefulness. It is by such investigations, in fact, that natural history is rendered practically useful, and is brought to bear upon the ordinary business of life. In expatiating upon the advantages attending the study of nature, we have already touched upon this subject, and have shown in what a variety of ways a slight knowledge of natural history might be turned to practical use. When we reflect how little has yet been done in ascertaining the chemical properties of animals and vegetables, there seems no valid reason for supposing that beneficial discoveries may not still be made, provided due attention be given to such enquiries. The properties of nature are inexhaustible; and man, as he advances in civilisation and refinement, acquires new desires and new wants. How astonished would the ancient inhabitants of the Scottish islands have been, had they foreseen that the loads of sea-weed called kelp, which they suffered to lie and rot upon their coasts as utterly valueless, would become a source of immense wealth, and that this manufactory would suddenly be destroyed, by the discovery of cheaper and better substitutes.

(123.) 3. We come now to the third head of those

enquiries which concern the natural history of an animal; namely, its relation to other beings. The very dissimilar forms which nature has given to most of the great divisions of the animal world are so striking in themselves, that the more general relationships are obvious to common observers. Thus, a quadruped, a bird, a fish, or an insect, is known, in ordinary cases, at first sight. Even if we descend to more particulars, and proceed to assort quadrupeds, for instance, into separate divisions, we see plainly that a lion has no affinity with an ox, or a monkey with a mouse, further than, as being quadrupeds, they have a greater relation to each other than with birds, fishes, or insects. Relations, therefore, are either general or particular; but both terms are used comparatively. Thus we may say that the elephant is related to the ox, by being in the same natural order: but this relation, comparatively speaking, is only general; because, between these two quadrupeds, other forms or species intervene, which show a more particular resemblance to one or to the other. Hence we see that, as there are different degrees of relationships, it becomes necessary to give a more precise analysis of the term, and to ascertain in what manner these different degrees of relationship can be defined.

(124.) Relations or resemblances, in the ordinary acceptation of the words, have long been considered as of two kinds, expressed by the terms *analogy* and *affinity*. By the first, we understand an external resemblance or similitude to another object, which is nevertheless different in its form, structure, habits, or some other important circumstance: here the resemblance is consequently superficial. By *affinity*, on the other hand, we imply such a resemblance in those characters just mentioned, and such a strong similarity in the detail of the structure of two animals, that they are only kept distinct by a few peculiarities of secondary importance. These two sorts of relations have been apparent since men first began to reason on the things they saw; but although admirably explained by one of our modern zoologists, they have been so confounded and obscured by the writings of most others, that some, bewildered by the looseness of the existing definitions, have gone so far as to deny their very existence. The following illustration, however, will render the distinctions here given, intelligible to the most unscientific reader. Let us compare, for this purpose, the full-bottomed monkey, or the *Colobus polycomas* of Geoffroy, with the African lion (*Leo Africanus* Sw.), and we are struck, at the first glance, with their mutual resemblance: both have long manes, hanging over their shoulders; both have a slender tail ending in a tuft of hair; and both have the fur, in all other parts, short and compact. Had we no knowledge that such a monkey really existed, and merely saw its figure, we might be tempted to think it was a bad representation of the lion. Strong, however, as this resemblance, at first sight, undoubtedly is, we soon discover it is merely superficial. It is essentially, in fact, a monkey in the garb of a lion; without possessing any thing of the characteristic structure, the habits, or the economy of that quadruped which it represents: the relationship, in short, is one of analogy only; for, were it of absolute affinity, they should be classed together: both would then be ferocious, and would possess that particular structure peculiar to carnivorous quadrupeds. This, therefore, is an instance of *analogy*. Let us now look to one of *affinity*. The lion and the tiger, although by no means so alike in their external aspect as the last, are yet known by every one to be closely allied. They resemble each other, not only in their manners and external organisation; but both possess that peculiar conformation of teeth, claws, and of internal structure, suited to their carnivorous nature. Their difference is almost confined to their external aspect; whereas, in the former case, the external aspect constitutes the only point of resemblance. While speaking of the tiger, we may mention another instance of analogy equally striking. Nature seems to delight in showing us glimpses of that beautiful and consistent plan upon which she has worked, by giving us a few instances of symbolical or analogical representations, so striking and unaswerable in themselves, that they are perceived and acknowledged by all. What, for instance, can be more perfect than the analogy between the Bengal tiger and the African zebra? both of them striped in so peculiar a manner as to be unlike all other quadrupeds, and both so perfectly wild and untameable as to have resisted every effort employed for their domestication. No one, however, would proceed, upon such grounds alone, to class them together: for the one has the habits of a horse, and feeds upon herbs; while the other, like the lion, devours flesh. The preponderance of characters, in short, denotes their respective affinities; while their analogical similitudes are drawn from those left in the lighter scale. Nor are such relations confined to one class, or to one division, of animals; for the further the student proceeds, the more universally can he trace them throughout nature.

(125.) It is a common and a just comparison, to liken the vulture and the eagle to the lion; the two first being among birds, what the latter is among quadrupeds, — the tyrants of their respective races.

> "The eagle he is lord above,
> The lion lord below."

This comparison, moreover, is rendered doubly accurate by a singular analogy of structure, which, as we do not remember to have seen it noticed, may be here advantageously introduced. The lion, — apparently to prevent the adhesion and drying of fragments of his bloody meal upon his skin, where it might putrefy and create sores, — is provided with a bushy mane, which prevents the blood or gore from coming into immediate contact with his skin, and which he can thus shake off with ease. Now, if we look to the greatest number of the vultures, we find that nature, to effect the same purpose, has given to them a similar provision. *They* also have a mane upon their neck; not, indeed, of hairs, but of feathers longer than the others, and generally so stiff and glossy, that any substance which may come upon them can be shaken off with ease. The vulture is, then, the lion among birds; and affords one of the thousand proofs, that relations of analogy can be found in animals of different classes, no less than between others more closely related. From these proofs, which come home to the conviction of all, the student will readily perceive that there are relations of analogy, as well as relations of affinity; and he will plainly see the theoretical difference between them, disconnected from any particular system or theory. To deny the existence of such relations, is to deny the existence of our senses.

(126.) It further appears, from the examples just given, that there are different degrees of analogies; some being more striking than others: hence they become either *immediate* or *remote*. We say that an analogical resemblance is *immediate*, when it concerns animals of the same class, as that of the monkey and the lion; and we term it *remote*, when the comparison is made between individuals of different classes,— between quadrupeds and birds,— as just exemplified. In the former case, the animals compared come nearer to each other in the order of nature than do the latter, and their mutual resemblance is consequently greater. The degrees of affinity, on the contrary, are much fewer, and more circumscribed in their range. No animal can have an affinity, except to those which stand in the same group, or which immediately precede, or immediately follow it. But its analogies, as will hereafter be seen, may be traced throughout all other groups of the same class, and even, in some cases, throughout the whole animal kingdom.

(127.) It must, then, be received as an incontestible truth, that every animal has a twofold relation to others. By one of these it is united, like the link of a chain, by direct affinity to others of its kind:

while, by the second relation, it becomes a type or *emblem* of other animals with which it has no positive connection, or consanguinity.

(128.) Having now briefly stated, in as comprehensive terms as the intricacy of the subject will admit, the theoretical difference between analogy and affinity, and given the student a leading clue by which he can separate the diversified relations he will find in nature, we shall, in the succeeding chapter, conduct him a step further; and, by endeavouring to point out such considerations as should influence all natural arrangements, or theories of classification, prepare him, in some measure, for entering upon the philosophic investigation of nature.

CHAP. III.

ON ARRANGEMENTS GENERALLY; AND ON THOSE CONSIDERATIONS WHICH SHOULD FORM THE BASIS OF EVERY ATTEMPT TO CLASSIFY OBJECTS ACCORDING TO THE SYSTEM OF NATURE.

(129.) THE innumerable objects composing the animal world, may be compared to the isolated facts of all physical sciences. For unless they are arranged and digested under proper heads, no general conclusions from them can be drawn. No sooner, therefore, has the naturalist become acquainted with the forms of the objects he studies, than he proceeds to arrange them according to their agreements and disagreements. He first places them in primary groups, as an entomologist would separate the beetles from the butterflies; and these, from the bees and the flies: from each of these, again, he proceeds to make other divisions; separating the butterflies which fly by day, from those which are nocturnal, and so on. This is arrangement or classification; from which all systems or methods originate. Now, it is obvious, that if we are not guided in this proceeding by some general rules known to be universally applicable, every one may consider himself qualified to follow his own impressions, and to make that arrangement which he thinks best. Hence have originated the innumerable systems and methods which have been, and are still, in use. One writer attaches a primary importance to particular characters, which another undervalues; a third rejects both these, and founds his system upon certain points of structure on which his predecessors have placed no value; a fourth, disregarding all outward organisation, builds his method upon internal anatomy. The first question, therefore, which a student naturally asks, is this;— Where, amid these opposing systems, am I to choose? None of them rest on, or appeal to, any general laws of arrangement, applicable to other departments of nature besides that upon which they treat. The classification of each author rests solely upon his own opinion how certain facts are to be arranged. Individual dogma seems to be the only basis. Mr. A. considers insects should be classed according to their wings; Mr. B. contends that they are best arranged by regarding their feet. What, then, we must first enquire, are those considerations which should guide us in a choice of system?

(130.) Now, systems may be of two kinds, artificial and natural. By artificial systems is to be understood any mode of arranging objects according to the absence or presence of certain given characters, without regard to such others as they may possess; or, if we arranged them simply according to their modes of life, without any reference to their particular structure. Thus, if all molluscous animals were arranged into those which had shells, and those which had none; and these former, again, into univalves, bivalves, and multivalves; this would be an artificial arrangement — because, by selecting these characters alone, and passing over every other as inferior, we bring animals together of totally different organisations. A natural system, on the other hand, aims at exhibiting that series which appears most to accord with the order of nature. It does not attempt to define groups so rigidly as to render them absolute divisions; but, by passing over solitary exceptions, rather seeks to gain general results, and to develope that uniformity of plan, upon which every object in nature was originally created. There can be, of course, but, one true natural system. But we may safely speak of all such as we have last defined in the plural number, because they all aim at the natural classification; whereas the object of an artificial system is merely to assist us in finding the names and properties of species. As we shall have occasion, hereafter, to treat of systems more at large, it is sufficient, for our present purpose, merely to give the student a general idea of their respective natures.

(131.) With his materials before him, in the shape of notes and specimens, the young naturalist is now to choose whether he will adopt an artificial, or aim at a natural, classification: that is to say, whether he will learn the names of objects by rote, as he would learn the words of a dictionary; or whether he will try to combine his objects in such a way as to discover the principles upon which their variations are regulated. By choosing the latter plan, — to pursue the simile, — he will endeavour to dispose the words of his dictionary in such order, as that they may produce harmonious sentences, or intelligible truths. It is plain, that the philosophy of natural history is entirely confined to such systems as are founded on the considerations last stated; for no pursuit deserves the name of science, strictly so termed, which seeks not to obtain general results, or to investigate and develope general laws. There is nothing very intellectual in simply investigating the form of an animal, and in recording its manners; because, in these matters, we merely confine ourselves to objects of sense and sight. To frame a good artificial system, however, is proceeding a step beyond this; because, to make judicious combinations, easy to be understood, requires a peculiar tact, and no small acquaintance with the different forms of nature. On this account, it may be as well, perhaps, to enumerate, generally, what are the advantages, and what the disadvantages, of *artificial* systems, before we make the same enquiry into those we term *natural*.

(132.) Artificial systems, then, upon the first view, appear more calculated to facilitate our search after an unknown object, than any other mode of classification. From merely directing the attention of the student to one or more striking points of structure, they convey to his mind an idea of *simplicity* which is at all times captivating, and which, to the young beginner, is particularly inviting, from the impression it gives of a diminution of labour. This impression is generally well founded; for it is obvious, that the more numerous are the characters employed, the greater is the trouble imposed upon the student, and the more complex will be the system they are in. For instance, if he be an ornithologist, and adopts an artificial arrangement, he finds that all birds are divided into two large groups — land and water birds — the distinctions of which, in ordinary cases, are immediately comprehended: but if he prefers a natural system, he has to peruse the characters of five or more primary groups, before he can refer his subject to one of these primary divisions. Should he, again, wish to understand the name of one of those soft slimy marine animals destitute of a shell, and of which nearly the whole of the Linnæan class of *Vermes*, or worms, is composed, his labour will be still more abridged by using an artificial system. He turns to the *Systema Natura*, and he immediately finds that this animal will come under the order of *Mollusca*, concisely defined as " naked simple animals, not included in a shell, but furnished with limbs." Here, then, he looks no further, but proceeds at once to ascertain the genus, and possibly the species. Should he, however, wish to ascertain the natural group of his subject, his trouble is increased tenfold. He must first ascertain to which of the three great classes of animals, — the *Radiata*, the *Annulosa*, and the *Mollusca*, it really belongs: and this, as the science now stands, will oblige him, in many cases, to dissect his subject; because each of these classes contains " naked simple animals, not included in a shell, but furnished with limbs." From these examples, sufficient to illustrate the simplicity of a good artificial system, it will be immediately perceived how much the labour of any one who searches after names only, is abridged by the one method, and increased by the other. The truth is, that the perfection of an artificial system consists in making us acquainted with an object by the shortest and most easy way possible; and that all idea of following the order of nature should be totally abandoned, as inconsistent with this primary object. It is one of the greatest merits of Linnæus —who knew, better than most of those who have come after him, the true difference between the systems in question,—that he saw the truth of this position, and acted up to it. His primary object was to make things known by their names in the most easy manner; and he had the sagacity to foresee that, by this plan, he should win over to the study of nature, numbers of ordinary minds who would otherwise have regarded it as too intricate and difficult; and he succeeded to admiration. The world was astonished at the simplicity of his system; and delighted to find they could ascertain, with so little study, the scientific name of an animal or a plant in that book which was looked upon as the mirror of nature,—as, to a certain degree, it really was. To blame this great genius, therefore, for his unnatural combinations, is to blame him for what may almost be termed one of his greatest merits. He knew that his system, in many parts, made some approach to what he imagined was the system of nature; but he also knew, that to attempt following this up into all the details of his work, would be altogether premature, if not impossible. All this, we repeat, he well knew, and he framed his system accordingly. The perfection, in short, of an artificial arrangement is, that it should be *thoroughly* artificial: the divisions, as far as possible, should be made *absolute;* and no affinities, however natural, or however palpable, should be suffered to interfere or stand in the way of this primary object. We would almost say, that for amateurs, or for those who merely seek to know scientific names, a thoroughly good artificial system is the best *for use*. To judge from present appearances, natural history, as a science, is fast approaching to that state when its cultivation will be confined to the man of leisure and of learning ; to those who are installed in the precincts of a public museum; or who are possessed of a library and collections which would cost a fortune to purchase, or a lifetime to acquire. That the science should be daily becoming more difficult, is not to be wondered at, or to be regretted, because an accession of new objects calls for greater labour of investigation; and no one can lament the extension of knowledge, however he may be thereby prevented from acquiring it himself. If there is any ground, therefore, upon which we can advocate the expediency of a good artificial system, even in these days, it is that of enticing over to the admiration of nature, those persons who, in the present state of the science, are frightened at its difficulties, or turn away in disgust at the dry uninviting manner in which nature has been enshrouded with scientific technicalities. Such persons, it is true, would not themselves, by adhering to an artificial system, do any thing to develope the philosophy of the science, but they might enrich its records with innumerable facts, which might be stored for future use ; and even they themselves might in time be converts to a more just mode of pursuing the study. Perhaps the best artificial systems of modern times, are those proposed by such writers as divide every group into two; the one having *positive*, the other *negative* characters. These, indeed, are so simple, that the most illiterate can understand them; for we have only to see what an animal *has*, and what it *has not*, to find it out and determine its name. Such methods of arrangement, as might have been expected, violate the series of nature at almost every step; but this, as before observed, is of no sort of consequence in a really artificial system, where the primary object is to arrange animals, as nearly as it is possible, on the same plan as words are placed in a dictionary.

(133.) The disadvantages, however, of all such methods more than counterbalance the facilities they appear to offer. In the first place, they must necessarily disregard the order of nature, which it is the chief object of this science to discover and to unfold. As their perfection consists in their absoluteness, they must separate into widely different groups, animals which are not only of the same genus, but actually of the same species. For instance, no arrangement of insects appears more simple, and even in some respects more natural, than that which divides them into such as have wings, and such as have none. Yet if this plan is so rigorously acted upon, as to render it a correct guide or index to the nomenclature of insects, we must place the female glowworm in one division, and the male in the other; the first being without wings, while the latter has four, two of which form cases for the protection of the others. The sexes of several moths, where the same singular differences are found, must, on the same principle, be likewise separated. It might be easy, indeed, in an artificial system, to place such apterous insects in a division by themselves; but what mistaken ideas would such a plan give rise to! To render such insects intelligible, they must have a name; and we should either be compelled to introduce the same genus and the same species into two different divisions — perhaps volumes, — or we must call the male by one name, and the female by another! Besides this, as artificial systems are framed upon no *general principles* of classification, no stability whatever is given to the very elements of the science; or rather, there is an absence of all elements. As there can be neither science nor philosophy in an alphabetical arrangement of words, so there can be none in a system of animals framed only with a view of making them easily found out. Simplicity, also, which seems at first so captivating a feature in such methods, is more superficial than real. Do what we can towards defining groups so strictly that no exceptions shall occur, and no deviations be found at variance with our generic characters, still we shall soon discover how impossible it is to circumscribe nature, even in her lowest groups: we shall constantly be meeting some species which depart from our arbitrary standard of character, and which oblige us to make new divisions for their reception: these divisions will finally be so multiplied, and so intricate, that our artificial method will become more complicated than the most elaborate natural system: its simplicity, in fact, will be destroyed; and we shall lose as much time in becoming acquainted with a superficial classification, as would have sufficed for the acquirement of a sound one.

(134.) The glaring violations of nature which result from a strictly artificial system, have been felt so forcibly by nearly all the best systematic writers, that they have endeavoured to unite facility of research with some attention to the order of nature; hence the origin of mixed systems, such as that of the *Règne Animal* of Cuvier, and the *Genera Insectorum* of Latreille. Arrangements of this description have been, and still are, highly useful; inasmuch as they bring together the scattered fragments of the natural series, disjointed and severed by artificial methods of arrangement. Yet they are not so useful to the searcher after species, because a wider latitude is given to the definitions; nor can they be considered as built upon philosophic principles, for they commence upon no universal and acknowledged truths of natural classification. They frequently bring together natural groups; but after proceeding awhile in the evident order of nature, they suddenly stop, and enter upon another portion of their subject, as if it had no connection whatever with that which they had just left. With this they go on in some regularity; but soon another interruption is apparent, another gap is to be leaped, and another series is begun upon, as if it had nothing to do with the last. These systems, which exhibit nature, not as a whole, but as pieces, may be compared to fragments of a chain, each composed of an unequal number of links, which as far as they extend are perfect, but whose two extremities show the marks of being violently dissevered from other portions. They are — to use a homely expression — bits and scraps of that which is, naturally, a uniform and connected whole. To illustrate this, we need only advert to the best classification of quadrupeds now extant. Commencing with the orang otan, the series passes from them to the baboons, the monkeys, the howling apes, the prehensile monkeys, the lories, and the bats. So far there is an evident appearance of a natural series, and we begin to think the author is really arranging animals *according to their organisation*; but we have arrived at the end of the first fragment of the chain, and, dismissing all idea of continuity, we are to begin on another. Immediately after the bats are placed the hedgehogs, and following these come the bears. Every person, possessing the slightest knowledge of these animals, at once perceives how unnaturally they are thus combined; and when he learns that there is no other reason for this, than because they happen to agree in some one or two points of organisation, arbitrarily fixed upon as the groundwork of the system, he may fairly question whether such a series exhibits the true order of nature. Mixed systems, moreover, lie under the same objection which has already been urged against artificial ones — that is, they exhibit none of that *harmony of plan* between the different groups, which must necessarily form a part of the system of nature. Nor do they even show an uniformity in the minor divisions of that particular department upon which they treat. An arrangement of quadrupeds, for instance, is made, as if quadrupeds had no reference to birds, or as if each had been created upon distinct plans. So in regard to fish, or reptiles, or insects; each is successively arranged independently of the others, as if they were so many isolated systems, and not merely portions of one. We may, perhaps, be censured for dwelling upon the defects of some of the most influential authorities of the day; but it should be remembered, that error must be removed before truth can be established. Natural history, like all other human knowledge, is progressive; nor does it at all follow, that, because our predecessors may have been mistaken on some points, we are to set aside their authority, or undervalue their labours, upon others. It has been well said of such men, that even their errors are the errors of genius; and that they are calculated, if rightly used, to teach wisdom to such as come after them. Let it be remembered, also, that we are now investigating natural history as we would do any other of the physical sciences; and that we can only hope to advance its interests by making it subject to the same general principles, and the same rules of investigation, which are applicable to all. There cannot be a doubt, that mixed systems, however objectionable upon the grounds we have stated, have done incalculable good, and have brought the science to such an advanced state, that, in these latter days, a glimpse of the natural system has at length opened upon us.

(135.) We shall now shortly consider the nature of those classifications which aim at exhibiting uniform principles and general results, and which we have consequently termed *natural* systems. The term, we before observed, is in some respects objectionable, because, without the explanation already given (129.), it would seem to imply a plurality of natural systems; whereas, in fact, there can be only *one*. It is not to be supposed, however, that all the laws of natural arrangement are to be developed at once; or that, amid the infinite diversity of resemblances which we see in the animal world, erroneous combinations may not be formed, which will nevertheless wear the *appearance* of following nature. Hence arises the necessity of discussing more at large the nature of theories, and the considerations by which they are to be verified. For our present purpose, however, it is merely necessary to state, that a natural system of classification aims at two primary objects: first, the arrangement of all objects according to the scale or series which they may be supposed to hold in the order of nature; and secondly, to discover, from such an arrangement, the general principles which govern their variation, their structure, and their habits. The first of these objects is likewise aimed at by the mixed methods of classification just noticed; but the latter—that is, the discovery of general laws, or of the fundamental elements of the science—is the peculiar characteristic of natural systems; because they, and they alone, endeavour to solve the principles of those harmonies and connections, which, reasoning from analogy, we feel convinced must be regulated by definite laws.

CHAP. IV.

ON THEORIES IN GENERAL; AND ON THE MODES AND CONSIDERATIONS BY WHICH THEY ARE TO BE VERIFIED.

(136.) It has been shown, in the preceding chapter, that there are three modes by which the objects of nature may be classified; and that one of these — that is, the natural system — is alone conducive to the advancement of natural history as a physical science. To this, therefore, we shall hereafter confine our attention; because the principles of *this* science must be discovered by a similar series of inductive generalisations to those used in every department of natural philosophy, " through which one spirit reigns, and one method of enquiry applies."

(137.) Let us suppose, then, that an entomological student, with a well-filled cabinet of unarranged insects, having his mind well stored with those simple facts regarding their structure and economy which he is to look upon as solid data — let us suppose him to commence the arrangement of the objects before him, according to what he thinks their true affinities, and with a view of verifying or discovering their natural arrangement. He commences by placing, one after the other, those species which bear the greatest mutual resemblance; and for a time he proceeds so satisfactorily, — he finds the several links of the chain, as it were, fit into each other so harmoniously, — that he begins to think the task much easier than he at first expected; and that he will not only be able to prove, by these very examples before him, the absolute connection of one given genus to another, but also to demonstrate that the scale of nature is simple — that is, passing in a straight line from the highest to the lowest organised forms. All these ideas, however (generally resulting from partial reasoning or from limited information), are soon found to be fallacious. As the student proceeds, he meets with some insects which disturb the regularity of his series, and with others which he knows not where to place. He still goes on, however, introducing the former, in the best way he can, among those to which they have an evident affinity, and placing the latter by themselves, under the hope of finally discovering their proper place. The further he proceeds, however, these difficulties are rather increased than diminished. He remodels his groups, and alters his series; still he cannot reduce all into harmonious order. What he gains by one modification of arrangement, he loses by another; and affinities, which were preserved in his first series, are destroyed, that a place may be found for other insects, which seem to have equally strong relations, although, in some respects, they evidently disturb the order of progression. But his difficulties do not terminate here; for, admitting the possibility of his success in bringing every species into an appropriate group, the union of these groups among themselves opens a new source of embarrassment. It is plain that, in the order of nature, they must follow one another in some sort; for if there were no *progression* of developement, all animals would be equally perfect—that is to say, have the same complexity of structure. Here, then, lies his difficulty. He perceives, perhaps, an evident affinity between two groups, by species which seem to blend them together, and to conduct him, by an almost insensible gradation, from one to the other. He therefore concludes this to be the natural series, and he approximates them accordingly: presently, however, upon looking more attentively to his other unsorted groups, he finds not only one, but several, each of which, in some way or other, shows an approximation just as close to his first group, as that does which he has previously made to follow it; and he is as much at a loss how to dispose his groups in natural succession, as he was how to place the species which they contain. The same results also attend his attempts at improving his arrangement of groups: what is gained by shifting one so as to follow another, is lost by dissevering it from that with which it was previously united: until, with all his assiduity and trials, he finds there is still a remnant of " unknown things," which stand disconnected, as it were, from the series he has formed; and which cannot be made to fall into place by any contrivance he can devise.

(138.) Now, the first question which arises in such a state of things—a state which every naturalist has repeatedly experienced,—is this;—What is the series of nature? Is it simple, or complex? and in what manner, or by what rules, am I to distinguish the different natures of all these complicated relationships or resemblances, so as to determine which is the natural series of the groups now before me? Here, then, commences the philosophy of the science. For we have either to determine these questions by a long process of inductive generalisations, which will probably occupy years of incessant study; or we must have recourse to the experience of others, and proceed to verify, by the subjects before us, those general conclusions which others have arrived at upon the points in question. This, therefore, will be the place for giving to each of these general conclusions some consideration.

(139.) It was long the opinion of philosophers, that the chain of being, or, in other words, the order of nature, was simple. So that, between man, and the minutest animalcule invisible to the naked eye, there was an innumerable multitude of organised beings descending imperceptibly in the scale, and forming a simple continuous series, like the links of a chain; the first of which was very large, and the latter very small, the intervening ones gradually lessening as they approached the lowest extremity. Now, this theory has long since been abandoned; because, although we can select from the animal world a series which will answer to such a theory, we should still be obliged to omit nearly one third of the animals already known, which will not, by any possible contrivance, fall into a linear series, and which consequently demonstrates its fallacy. The very instance we have just given, makes this apparent to the most inexperienced student. If the chain of being had been simple and linear, he would have had no difficulty in placing his insects in such a series; and one would have followed the other, with only such intervals as future acquisitions or discoveries might be supposed to fill up. At all events, he would not have been perplexed by an apparent multiplicity of relations, branching off in different directions, and totally discomposing his linear series. Every object which is arranged, like the links of a chain, in a simple line of progression, *can* have but two *immediate* affinities: one, by which it is connected to that which precedes it; the other, to that which follows it. The student, therefore, at the very commencement of his study, has a demonstrative illustration that the chain of being is continuous, yet at the same time *not* simple. This truth being verified, he has next to enquire in what mode this continuity is preserved, and what is the actual course it takes in its progress from the most perfect to the most imperfect organised beings.

(140.) Now, to solve this latter question, there are, as it has been justly observed of natural phenomena in general*, three modes by which we may proceed. First, by inductive reasoning: that is, by commencing with the lowest or nearest approximations, as that of species to species; forming groups of them, and then endeavouring to discover the *degrees* of affinity or of proximation which these groups bear to one another. Secondly, by forming at once a bold hypothesis, particularising the law, and trying the truth of it by following out its consequences, and comparing them with facts: or, thirdly, by a process partaking of both these, and

---

* Herschel, Dis. (Cab. Cyc. vol. xiv.) p. 198.

combining the advantages of both without their defects; viz. by assuming, indeed, the laws we would discover, but altering and modifying them in the process of their application, so much as to make them agree with incontrovertible facts.

(141.) Of these three modes of investigation, the first and the last are more adapted to ordinary capacities than the second; because to conceive a bold and comprehensive theory, which should carry with it a semblance of reconciling, and reducing to general laws, a multitude of facts apparently anomalous, requires a proficiency in science which few have the talent or the means to attain. This objection is applicable also, although in a less degree, to the third mode of investigation; for here also, as we are to assume certain laws, the assumption, — in order to wear any appearance of truth, or to raise in our minds any solid hope of success in working it out, — must be the result of much experience and of extensive research. He, therefore, who would proceed with that caution so necessary in the intricate path he has now entered upon, should either begin his ascent at the very lowest steps, and never venture forward until he has obtained a sure footing upon that; o. he must trust himself to the guidance, in the first stages of his journey, of those who are familiar with the road, and have already affixed certain landmarks sufficient to point out the direction he is to pursue. But to drop metaphor; the student must proceed on one of the following plans: — He must either commence, as pointed out in the first of these methods, by supposing no general laws have yet been discovered, and that he may possibly find a clue to their developement; or he must begin by assuming as true, those laws which have been demonstrated by others, and proceed at once to verify them in the groups he is about to investigate.

(142.) So little had the philosophy of zoology been attended to by those who, nevertheless, in other respects, have been the greatest benefactors to the science, that it is only within the last fifteen years we can date the commencement of such enquiries. It was then that the first efforts were made to reduce it to an inductive science, to be prosecuted by the same method of enquiry which had long been employed in other departments of natural philosophy. Hence it is, that *this* science is based upon fewer known and acknowledged truths than any other. So strong has been the force of prejudice in favour of artificial methods, and so little disposed are the naturalists of the old school to quit the beaten path they hitherto traversed, that if it were asked, what were the number of general laws or inductive generalisations of the highest order, at present admitted in this science, we should be obliged to confess that only *one*, as yet, has been extensively verified. This *one*, however, is of the most comprehensive nature; since it regards the *chain of being*, or the order of succession, in the forms of nature which we are at present discussing. The law in question is this; — That the progression of every natural series is in a circle; so that, strictly speaking, it possesses neither a definite beginning nor a definite end; the two extremes blending into each other so harmoniously, that, when united, no marked interval of separation is discovered.

(143.) Of this law we have a familiar and a very beautiful illustration, in the annual revolution of the seasons; the months of which may be compared to a series of beings following each other in close affinity. If we begin with January, we trace the gradual developement, first of spring, and then of summer; from thence we pass into autumn; this season, again, melts into the winter of December; and we thus arrive again at the point from which we first set out. So, likewise, is a natural series of animals. If we begin at any given species or group, and trace its connection to others, we find, that, after being conducted through various modifications of the original type, we are insensibly brought to that type again; just as if, by passing the point of a pin over the figure of a circle, we should assuredly end where we began. Now, as this hypothesis has been amply verified by facts drawn from the animal and the vegetable kingdoms, it has assumed the character and the authority of a general law, and gives us no further occasion to seek upon what principle the series of nature is constructed. We shall have occasion, hereafter, to dwell more particularly on the comprehensiveness of this law, and the beauty of its application throughout nature: at present, we merely point it out to the student as that basis upon which all his combinations must be built, and as a fixed and determinate point from whence he may safely begin his journey onward.

(144.) The lowest combinations of objects, wherein this law can be traced, are those groups of species which were formerly denominated genera, but which are called by the moderns *sub-genera*,—a term indicative of their subordinate rank. The first process of generalising, or, in other words, the first stage of induction, is to bring an indefinite number of species into a group, which shall be so rigidly restricted, that little other variation is seen, among the individual species so associated, but such as arises from size, colour, or the greater or less developement of the same parts and the same organs. It is quite immaterial to our present purpose, whether we call these groups genera or sub-genera; but it is of the first consequence, that naturalists should agree in the meaning of certain terms or words. That such groups as we have just described are *natural*, can admit of no doubt. The olives, the cones, and the cowries among the *Testacea*, are good examples; while, in entomology, we have the white garden butterflies (*Pieris*, Lat.), the blues (*Polyommatus*), the coppers (*Lycæna*), the hair-streaks (*Thecla*, Lat.), and the rove beetles (*Cicindela* Lin.). In the British species of all these respective groups, we have a perfect illustration of the above definition of a sub-genus, and of that degree of variation which is found among species so grouped. Hence it follows, generally speaking, that the determination of a sub-genus (or of a group so denominated) is one of the most easy things imaginable; for, whenever we meet with a species which shows a marked affinity to any one of the above assemblages, yet possesses a peculiarity of structure which those have not, or wants another which they possess, we may in ninety-nine cases in a hundred conclude it to be the type of a sub-genus, which future discovery will most probably augment by other examples, and which higher degrees of induction will

P

show us is absolutely essential to the harmony of the whole.

(145.) The student must not, however, suppose that all sub-genera are so comprehensive, or so readily detected, as those which, for the sake of strong examples, we have just instanced: very many, so far as we yet know, are composed but of one species; and, generally speaking, the number of species is small. He must not, therefore, be apprehensive he is carrying the above theory too far, when, in the arrangement of his collection, he places at short intervals of separation, many insects, as probable types of sub-genera, of which he has but one example. His fears, that he is making needless divisions, may be quieted by two considerations: first, that natural groups do not depend on their numerical amount of species; and, secondly, from the amazing number of nature's productions already known, and of which he has not, probably, seen one tenth part of such as actually exist in collections, setting aside those which have not yet been discovered. Even admitting that his collection is very extensive, and that there is consequently a greater chance of his finding more than a solitary example of a supposed sub-genus, still he will frequently be deceived in his estimation of its extent, which can often only be learned from books, or from an extensive acquaintance with the contents of other cabinets. A singular and striking instance of this is afforded by the sub-genus *Eudamus*, first defined by us (*Zool. Ill.* 2. pl. 41.), and composed entirely of the swallow-tailed skipper butterflies of the family *Hesperidæ*. Only one species was known to Linnæus;

and so few were since added, that not one of the modern entomologists ventured to arrange them as a distinct group. It happened, however, that we took a predilection for these butterflies when in Brazil, and collected them with great assiduity: the result is, that no less than eighteen species were found in that limited portion of the country which we explored. We are acquainted with several others, and new ones are still coming to light; so that it is very probable, in a few years, that this single sub-genus, not more than one or two species of which are usually seen in collections, will comprise fifty. We mention this instance, not as encouraging the student to increase the present overwhelming list of sub-genera — as too many are now doing — but as an example how impossible it is, in general, to judge of the real numerical contents of a natural group, from the examples usually seen in cabinets, or even from the species that have already been described.

(146.) The second stage of generalisation is to ascend from sub-genera to genera, or, in other words, to combine an indefinite number of those first, or lowest assemblages of species, just described, into a group of the next rank or denomination. Now, this group we name a genus. The question therefore is, whether there is any rule as to the specific number of sub-genera which naturally constitute a genus, or is this number indefinite, depending entirely on the greater or less variety of forms, which show a common tendency to unite into a circular group superior to themselves?

(147.) Here, again, the progress of the enquirer

is arrested. He has set out with assuming as correct, one great law of nature, — the circular progression of affinities: but now he is to enquire, before he can proceed further, on what principles he is to connect his sub-genera, so as to preserve their affinities, and yet form them into assemblages of a higher order or superior value. The answer, theoretically, is obvious to every one accustomed to logical reasoning. If, in all natural groups, the progression of affinity is circular, then the contents of a genus, which is a natural group, must be circular also. Such is the application of the law in question; for it cannot be supposed that the higher divisions of nature, as classes and orders, should demonstrate this circularity, and that the other subordinate groups should not: this, were it true, would disprove a unity and consistency of plan, and the law, not being general, would be no law. A genus, then, to be natural, must only contain, of necessity, such sub-genera or minor assemblages of species as will, collectively, show a circular progression of affinity. Such is one of the requisites of a natural genus. Still the question of *numbers* is to be investigated. What are the natural divisions of a group? in other words, how many of those, here called sub-genera, constitute a genus?

(148.) Now, as we are proceeding analytically, under no assumed law but that which regards the circular theory of affinity, we must have recourse to observation. If we find that three, or four, or any other definite number of sub-genera, by being placed together, appear to form a circular group,

We may, in this stage of our enquiry, have some reason to suppose that it is a natural one, because it exemplifies the law in question.

(149.) It has been found, however, that this is not a sufficient verification of a natural series; and for this reason: in the infinite variety of animal forms, there are so many mutual resemblances, that if our only object is to arrange them in circles, we may combine them in different ways, all of which will wear a *primâ facie* appearance of being more or less circular; or if any unusual *hiatus* or gaps appear, we are immediately ready to smooth over the difficulty, by concluding that they do not really exist in nature, but only in the paucity of our materials. And we are the more inclined to yield to this persuasion, since naturalists universally admit that such intervals really do exist in nature, either from the extermination of some animals by man, or from the changes which our earth has undergone. Against this disposition, therefore, to smooth over discrepancies in our supposed circle, by attributing them to causes which may or may not be the true ones, the naturalist must frequently contend. He may rest assured that a natural group can only have one series of variation; and if, by taking away some of its parts, and substituting others, he can still preserve its circular appearance, he has strong reason to doubt which disposition is the natural one.

(150.) Now, under these circumstances, the effective mode of proceeding is to form two or more of those combinations called genera, and then to compare their contents respectively with each and all.

Here we have first to bring into practical use the theoretical distinctions of analogy and affinity, already touched upon. If the contents of one genus appear to represent, in some few remarkable peculiarities, the contents of another; and if this can also be traced through a third or a fourth; we are immediately impressed with a conviction that this coincidence is the effect of *design*. It is clear that these resemblances, however strong, cannot be relations of *affinity*; because they occur in different circles, which would be broken up and destroyed, if these objects of resemblance were taken out and grouped by themselves. An example of this will best show its effect, and the violence it would commit on the law we have set out with. The most inexperienced ornithologist perceives a resemblance, more or less strong, between the cock, the wattle bird, the carunculated starling, the cassowary, and the wattled bee-eater. All these, in fact, at first sight, are immediately recognised by a head and face more or less naked, and ornamented with fleshy crests or wattles. Yet, if we concluded that this resemblance or relationship was one of *affinity*, and therefore proceeded to take all these birds out of the present groups they stand in, and place them in one by themselves, what a heterogeneous mixture should we have! Nor would this be all: the respective circles in which each of these types now stand, would of course be broken up, and another group would be formed of the wattled birds alone, which would be any thing but circular. Resemblances like these, as we known from experience, will be found in every natural group. When these groups are remote, we perceive, without investigation, that they are of analogy; but in proportion as groups approximate, other dissimilarities of course become less, so that when we descend to genera which follow or come very close to each other, it is impossible to decide, at first sight, whether the relationship be one of analogy or of affinity. But of this hereafter.

(151.) As every group, therefore, is found to contain some such striking modifications of form, it becomes necessary to ascertain how far they follow each other, in the same succession, in *each* group: for it is not to be supposed that they occur at random, or that they merely constitute a part of their own group, without having any uniform and definite station therein. The series of variation in one, must be the same in all. When, therefore, we wish to verify an assumed circle of affinity, our first business is to study the order of succession in which the subordinate forms in it occur, and then to compare it with other assumed circles. The proof that our arrangement of one is correct, is involved in the general verification of the whole. If the succession of forms in one and all of these circular groups agree, we can then have little or no doubt, that the order of nature has been discovered; for we shall then arrive at one general principle of variation, and shall be able to assign to each form the station it holds in its own group. If, on the other hand, we find no such analogy between the contents of our groups — if they contain no corresponding representations — and if they rest for their stability upon the mere appearance of being circular, — it is plain that there must be something wrong in our arrangement. It then becomes necessary to remodel the whole; and if, after this, no general results can be obtained—if there is no regularity in the occurrence of the same analogous forms in the different groups —our circles want verification, and must of course be considered hypothetical.

(152.) The presence of such remarkable forms as have been just instanced in a natural group, would seem to point out at once the most obvious means of deciding on the number of divisions which a genus contains; and as every distinct modification is the type of a sub-genus, we derive, in that first or lowest stage of induction which we are now supposing, great help in determining the number of divisions in a genus. It may so happen, that in one we reckon four, in another five, in a third seven, or even more. It then becomes an important question, whether these assumed divisions or types can be augmented or reduced so as to bring them to a definite number in each group. If this can be accomplished, it is clear that another principle of harmony will be discovered; and we shall have good reason to conclude that the number of divisions into which the majority of our groups can be divided, will be that most prevalent, if not universal, in all others. The verification, however, of such a theory cannot be satisfactorily attained until we quit genera, and ascend to higher generalisations; and for the following reasons.

(153.) Suppose, for instance, we looked to the genus *Trichius* among coleopterous insects, and agreed with an eminent entomologist in dividing it into seven principal sections or sub-genera; and we assume these to be natural. But on turning to the genus *Phanæus*, belonging to the same order, we find another celebrated writer declaring that, in *this*, all the species can be referred to five types: now the question is, how can these different opinions be verified, or made to agree? The rule, were the groups of a higher order, would be obvious. Do each of these divisions form circles of their own? if not, they are unnatural: but this test cannot be often applied to a genera; because it rarely happens that their sub-genera are so abundant in species as to form complete circles. Yet there are two modes which can still be resorted to, independent of any assumed theory, for ascertaining what is the determinate number in the groups before us. First, we should endeavour to see how far the seven divisions in one can be reduced to five, so that two of them are absorbed, as it were, into the others. If we find this to be impracticable, without destroying the equality of the divisions, we should reverse the experiment, and ascertain how far the five groups of *Phanæus* can be made into seven. If we succeed in this, or in the other, we establish an agreement; and, so far as we have then gone, there is presumptive evidence to favour the supposition that one or other of these numbers may be prevalent in other genera. The truth of such a theory, whether it be in favour of five, or seven, or any other definite number, depends on the extent to which it can be verified by observed or known facts. So that, although we may be able, as in the above instance, to make the divisions of two genera agree in their determinate number, and may therefore feel a disposition to build a theory upon such a coincidence, we must yet bear in mind that we have advanced but one step in the scale of induction; that it is not very difficult, even with a strict attention to the foregoing rules, to divide two genera, each into the same number of sections. To assume, on such slender premises, the existence of a general law, that all genera will be found similarly constituted, would be a total departure from that mode of enquiry which is absolutely essential to the prosecution of all physical science.

(154.) When, therefore, we have verified the prevalence of a definite number in the divisions of a genus, by comparing the contents of several, we may then advance a step further, and are at liberty, from the facts already elicited, to form a theory. It is essential, however, that, in so doing, we overstep not those inductions which lie before us, and which can be appealed to as instances of particular verification, and as presumptive evidences of the universality of the law assumed. By the process of investigation we are now pursuing, all deviations from the law we assume, must be accounted for on sound principles, or by the probable operations of known effects. Thus, for instance, many groups which, from having been already analysed and demonstrated, we know to be genera, contain but two or three divisions, others four, and many but one. Now, as this occasional paucity of forms in a genus may be accounted for by various natural causes (148.), we are not hastily to conclude that there is no definite number in nature, or that a genus may contain from one to twenty sub-genera, for aught we know to the contrary. Such imperfect groups, —

for so the wide intervals between the objects they contain proclaim them to be, — may be set aside for future analysis, when we have attained to higher degrees of inductive generalisation. It is those genera which, from containing numerous species and modifications of form, are usually termed *perfect*, which we are more especially to select as fit objects for the preceding line of enquiry. We all know, that the more numerous and varied are the materials given to us for accomplishing a given work, the greater will be the degree of accuracy attending the result, provided we use these materials for the purposes for which they were designed, and make each fit into the other with symmetry and order, so as to produce a perfect whole. Applying this to the question before us, we may safely assume that extensive genera are the most calculated to elicit the first principles of classification. They seem as if intended for natural storehouses, wherein we should find all sorts of implements with which we may try our hand at combining, changing, and remodelling, until we make all the parts, like those of a complicated puzzle, fit into each other; and produce, from what appeared a heterogeneous assemblage of isolated objects, a perfect tablet of order and beauty. It must be confessed, indeed, that such genera are the plague and torment of those who seek only to arrange them artificially; because the interchange of characters is so gradual, and the intervals between the more prominent types so filled up and crowded by connecting species, that it seems utterly impossible where to make one division begin, or where end. It is clear, however, that these are the very groups, above all others, wherein the principles of enquiry we are now recommending can be most successfully and most easily pursued; and for this purpose, as such, they should consequently be selected. When, therefore, we can draw any general deductions from the contents of several such groups, whether as regards the mode of variation in their subordinate forms, the characters of the forms themselves, or their definite number, we may rest assured of having committed no great error in their natural arrangement; and may safely assume the inductions thus obtained, as instruments to facilitate our further progress. Having now stated those primary considerations which appear necessary to determine, on sound principles of inductive science, the two lowest groups of nature, namely, sub-genera and genera, we may proceed a step further, and enquire into higher combinations.

(155.) The groups next in rank to genera, modern naturalists agree in calling *sub-families*. The name, however, has nothing to do with our present object, further than that it is necessary to give some designation to groups which are next in rank, or in comprehensiveness, to those last discussed. The determination, therefore, of a group of this sort,—no matter by what name we choose to call it,—must be regulated by the law we set out with assuming; that is, by the union of a certain number of genera, which, thus combined, produce a circular series. Here, again, the question of *numbers* arises. Now, bearing in mind, that the greater the degree of harmony and unity we can produce in our arrangement, the greater is the probability of our discovering the order of nature, it becomes essential to ascertain how far the laws regarding the combination of sub-genera into genera will assist us to combine genera into sub-families. We have supposed in the latter case, that the naturalist has found, in all his perfect genera, the prevalence of a determinate number of minor divisions; he is now, therefore, to try the strength of the law thence assumed, upon a more extensive scale. First, we must combine our genera in such a way that four, five, seven (or whatever the assumed number may be), make a circle of their own, more or less complete. We shall then have a certain number of circular groups, forming one of larger dimensions; and, proceeding in this way to form other assemblages of the same kind or rank, compare their respective contents. The first test of every such circle will be that its primary divisions or genera are also circular: the second, that these divisions or lesser circles, in regard to their *number*, are definite. If we find that their average number is greater or less in the sub-families, than in the genera, we must conclude one of two things; either that the number of types vary in different groups according to the rank or value of such groups, or that we have not yet discovered what is the true number most prevalent in nature. Every principle of sound reasoning is against the first of these suppositions; for if we suppose that natural groups are perfectly independent of any definite number of divisions, then (setting aside all experience to the contrary) we virtually deny uniformity of design in the details of nature, while we see and admit it in her grander features: besides, it is not to be supposed that such forms as we have elsewhere cited (150.), are scattered indiscriminately in their respective groups, without being accompanied by others, equally representing each other, and therefore implying, in the strongest possible manner, the existence of strict uniformity. We may, then, safely conclude, that if the number of our genera in a sub-family disagrees with the number of divisions in our genera, the fault lies with ourselves. We must again retrace our steps, perhaps abandon altogether the number first assumed as definite, and adopt some other more in unison with the facts before us. If, on the contrary, we can, in these new and higher groups, demonstrate the same prevalence of a determinate number, the strength of our theory is doubled. It has been well observed[*], that, " whatever error we commit in a single determination, it is highly improbable we should always err in the same way; so that, when we come to take an average, of a great number of determinations (unless there be some constant cause which gives a bias one way or the other), we cannot fail, at length, to attain a very near approximation to truth; and, even allowing a bias, to come much nearer to it than can fairly be expected from any single observation, liable to be influenced by the same bias." This useful and valuable property of the average of a great many observations—that it brings us nearer to the truth,—that is, to the determination of a prevalent number,—than any single

[*] Hersch. Discourse, p. 215.

observation can be relied on as doing, renders it the most certain resource in all physical enquiries, where the discovery of a general law is desired. If, for instance, we found, in ornithology, that twenty out of twenty-three sub-families, particularly abundant in species, could each be divided into seven groups or genera, and that each of these subordinate divisions was in itself circular, we should be justified in believing the determinate number to be *seven;* because the preponderance of evidence sanctions the conclusion, and leads us to believe that a more extended analysis of other groups will produce the same result. But if, in the remaining three, equally abundant in materials, we can by no possibility make out more than five circular divisions, we must either seek to equalise the results, or, if that fails, abandon our first theory, and commence anew. It will not be sufficient to argue that the two missing types of these groups *may* be supplied by future discoveries; because such a singular coincidence, of two missing types in each of three genera, carries on the face of it a high degree of improbability. It will be remembered, also, we are now supposing all the groups before us to be *perfect;* and, if perfect, then without any violent or palpable interruption in the line of continuity; in other words, presenting no interval, wherein, if these missing groups happened to be discovered, they could be naturally inserted. Nothing, indeed, can be easier than to start a theory on the universal prevalence of a determinate number, assumed upon the partial arrangement of one or two insignificant groups, and without complying with the conditions which authorise such groups to be called *natural*.

(156.) It must, nevertheless, be admitted, that groups so highly perfect as those we have just contemplated, are by no means of common occurrence; or, at least, our limited knowledge of nature has not yet enabled us to discover them. The most perfect group, in this sense of the term, in the whole circle of ornithology, is perhaps that of the sub-family *Picianæ*, or true woodpeckers, wherein we have ascertained, by the inductive process here explained, the circular succession of affinity in each genus, and consequently the characters of each sub-genus; all of which have actually been discovered, and are now in the European museums. Another natural group, even still more varied into different modifications, is that of the humming-birds (*Trochilidæ*); a group, moreover, which every one perceives is as natural as that of the parrots, the owls, or the birds of prey. The *Trochilidæ*, however, have not yet been analysed and grouped with that high degree of precision necessary to constitute a demonstration. The parrots, likewise, when we look to the diversity of their forms, may be included among the more perfect groups; and the ornithologist, really anxious to investigate truth, cannot have more favourable materials to work upon than these. There are so many considerations to be taken into the account, so many diversities of the same general structure not only to be reconciled but explained, so many degrees of relationship to be unravelled, and so many apparent anomalies to be illustrated by analogous examples in other groups, that a theory which explains these, must be considered as demonstrably true. Let the advocates for any determinate number, instead of declaiming in general terms in favour of their own opinion, and in abuse of others, throw aside such puerilities, as unworthy the name either of argument or of science ; let them in good earnest put their shoulders to the wheel, and resolutely sit down to study and develope the natural arrangement of any one of the groups just named : we shall then have a standard to which all parties can appeal ; we shall then see, beyond dispute, whether, in one of the most perfect groups in creation, nature has, or has not, regulated the variation of her forms by some definite number, or by some definite rule. It is a matter of perfect indifference to the man of true science whether that number be three, five, seven, or twenty. We want truth, and truth only ; and all that is true in physical science must repose on the experience or observation of facts within the reach of those who seek for them. One such analysis as we are now recommending, would tend more to the establishment of sound principles in natural history, than all the speculative declamation that was, or will be, ever written. It is surely not too much to expect such labour,—for labour it will assuredly prove,—from those who declaim against general views and particular theories, before they have informed themselves on the very first rules of judging which physical science imposes upon her votaries. Let us return, however, to the more immediate subject before us, viz. the verification of natural groups.

(157.) We have, in the last paragraph, spoken

more particularly of *perfect* groups; that is, of such as exhibit, in their circular progression, no wide or disproportionate gaps in their continuity. The naturalist, however, must not calculate on frequently falling upon a cluster of these, so near to each other, that every genus, for example, in a sub-family, shall be *perfect*. How, then, is he to proceed, since he cannot, in all instances, verify the law he has set out with assuming, *i. e.* that every natural group is circular? He must, in this dilemma, in the first instance, chiefly be guided by observation. Should he find that, by bringing together a certain number of groups, they will form a circle more or less complete, and of a higher denomination, he may, in the first instance, assume that the law in question has been complied with, if not in the letter, at least in the spirit. Some of the groups, thus united into one, may be *perfect;* whereas others may contain very few objects, and these objects, having distinct intervals between them, form *imperfect* groups; that is, they present such distinct and unequal spaces in the line of continuity, as to impress us with a conviction that intermediate forms are wanting, to render the circle perfect. Nay, it will sometimes happen that these last-mentioned groups contain but two or three individuals, while the others comprise forty or fifty. In cases like these, we must endeavour to discover how far these two or three individual forms,— placed together as the fragments, so to speak, of a circle,—are represented in the more perfect of the adjoining groups; and by the degree of continuity which these latter exhibit, estimate the extent of the hiati. If these isolated forms are represented in the adjoining groups (a fact which experience and critical examination alone will teach), then we have presumptive evidence for considering them as so many fragments or indications of a circle, the deficiencies of which we may form some idea of, by looking to the adjoining circles. To illustrate such a nice and somewhat abstruse subject more clearly, let us suppose there are two groups apparently following each other; one of which is perfect, and contains five principal variations of form; the other is imperfect, and contains but three, between which the intervals are of course much wider than between the other five. Now, if we are able to trace an analogous resemblance between three of one, and three of the other, we may fairly presume that the other two, which are deficient in the *imperfect* group, will, when discovered, exhibit a corresponding analogy. And we are thus not only justified in forming a theoretic notion on the nature of the forms of these missing types, but also in concluding those which we already have, to be parts of a distinct circle of their own, although its circularity is incomplete.

(158.) There is, indeed, one certain rule of deciding, in such cases, with almost mathematical precision, this is, by the law of representation; but to enter upon this subject at present would violate the main object we have endeavoured to keep in view. We are proceeding on that gradual mode of induction, which all who wish to understand or to benefit science must inevitably follow. We throw aside all theories, and assume nothing as granted but the circular progress of affinities. What has been already said, is applicable alike to all systems and all theories intended to develope the harmonies and relations of nature. We have, in short, studiously endeavoured to keep the mind of the naturalist unbiassed in favour of any system, and have restricted our observations to such considerations only as must be the foundation of all natural arrangement. But, as the admission of chasms in the order of nature appears to militate, at first sight, against the continuity, or rather the gradation, of forms in the creation, we may here make a slight digression on so important a subject. The most philosophic naturalist of modern times has placed this difficult subject in a light so clear and forcible, hat we cannot do better than condense, in one paragraph, his observations upon it, which are blended in the original * with other matters not adapted to this work.

(159.) The law of continuity, as it relates to forms of matter, may truly be proved possible in itself, and, in the next place, to exist in nature. Continuity in gradation of structure has, however, nothing to do with space or time. Matter, with respect to space, is capable of incontinuity; but with respect to gradation of form, it is as clearly capable of continuity. For this purpose, let us state a familiar case. Suppose a beautiful Grecian temple to be built in the neighbourhood of a sublime specimen of Gothic architecture. Let us further suppose, that between these two different buildings there is a transition made from one form to the other by an infinite

---

* M'Leay's Letter on Dichotomous Systems.

number of intermediate buildings, passing from the pure Grecian to the pure Gothic architecture. The continuity, whatsoever as to space the buildings intermediate in structure may occupy, will be perfect as far as relates to the gradation of form. And yet there must ever be some difference between the two structures nearest each other, in form: for if no interval exists, then these two must have the same structure, and one of them will thus produce no effect in continuing the chain of structure. In this kind of continuity, therefore, intervals between different forms are absolutely necessary; and if they do not exist, there is only one form. But in space or time an interval is impossible, and their continuity depends on this impossibility. On the other hand, continuity in gradation of structure depends on the existence of intervals; but requires, in order that the gradation be more distinct, that these intervals be extremely small and numerous. If only one mean be interposed between two extremes, there will be two chasms, but no saltus, and the three objects will be in continuity. Augment the number of various intermediate objects, and you only get the chasms more numerous, and the continuity more perfect. To argue, therefore, about the innate impossibility of the law, is absurd: the only question for us now to examine, being, whether such a continuity as I have described can be shown to exist in nature. I think I have proved this in my *Analysis and Synthesis of Petalocerous Coleoptera;* and what the Linnæans call natural genera, such as *Rosa* and *Erica*, are likewise all proofs of it: so that, if continuity manifestly holds good in these particular parts of the creation, which have been carefully examined, it may hold good in all. True it is, that nature does not always proceed *pari passu*. In the Linnæan genus *Psittacus*, — a group of very limited structure, — the chain is composed of a great number of links; whereas in *Pachydermata*,—a group presenting a very wide range of structure, — the number of links is comparatively small. Still there is continuity manifest in both; the difference depending merely on the relative distance between some two contiguous forms in each. Chasms in the chain may be numerous and small, as in *Psittacus;* or few and wide, as in *Pachydermata.*— Continuity in gradation of structure cannot exist, as we have seen, without intervals; and the size of these intervals does not lessen the truth of the chain, because some of the links may not yet be discovered. How, then, it may be asked, are we to prove that the chain is continuous? The reply is, simply by ascertaining which animals of one group come the nearest to those of the other. If there be no approximation — if all the animals remain equally distant—then there is no continuity; but if one animal of the one group approaches to the structure of the other, then there is a chain of continuity — possessing, indeed, but only one link, but not the less presenting a mode of transition from one form to the other. Thus, if the only animal existing between quadrupeds and fishes were one penguin, it would still be in the path of passage. But if a tortoise existed in addition, the chain would be more complete; and if one frog existed also, the chain would scarcely escape notice. In it, there is a regular and obvious gradation of structure: the chasms, indeed, remain vast; but there is no saltus, or leap, by nature, over one form to the other.

(160.) Continuity, then, as applied to the approximation or affinity of a series of animals, is not so expressive as the word *gradation*. And it is manifestly objectionable to employ, in science, the same term to express two very different meanings. After the preceding clear and able illustration by Mr. M'Leay, we need only touch upon the nature of those intervals or chasms therein alluded to, and which, to a certain degree, are absolutely essential to that diversity of structure we meet with in nature. When these spaces between two objects are very small, as among the parrots, they create no idea in the mind of an hiatus, or a manifest inequality or interruption of gradation, sufficiently wide for the insertion of other forms. But, when they are great, as between the different types of the *Pachydermata*, then they assume the character of chasms, which might be filled up by numerous other forms, calculated to make the gradation from one to the other more easy. It might at first be supposed, indeed, that this inequality of gradation either implies the loss of many links in the series of pachydermatous quadrupeds, or a want of due harmony and equality in the construction of nature's groups. A little reflection, however, on this apparent inconsistency will lead us to more correct conclusions. For this purpose we will still consider the *Psittacidæ* and the *Pachydermata* as striking examples of that inequality of gradation so frequent in different departments of the animal kingdom, and examine the question more closely.

(161.) The thick-skinned or pachydermatous tribe of quadrupeds comprise the genera of the elephant, rhinoceros, megatherium, and hippopotamus: these are well known as the most gigantic of all animals. We have ascertained, by analysis, that they form a circular group, and that the rank of this group is equivalent to that of a tribe. Yet, in regard to the number of objects it comprises, this is the most scanty tribe in the animal kingdom. It does not contain, in fact, as many individuals as are found in a single genus of parrots. Whence, therefore, arises this disparity? How are we to account for the wide intervals between the different *Pachydermata*, and the very small ones between the genera of parrots? To this we should answer, first, that many of these forms, which once existed, are lost; and, secondly, that their paucity, so far from disturbing the harmony and regularity of nature's system, tends to show it in a light directly the reverse. First, then, the extinction of numerous forms of *Pachydermata* rests on well known and incontrovertible facts. Not only are the fossil remains of hippopotami, of elephants, and of rhinoceroses, belonging to extinct species (and very probably to intermediate gradations of form), found in numerous and various parts of the world, and in considerable quantities, but modern geology has brought to light a whole family of these quadrupeds, represented by the megatherium, which are now so completely exterminated from the earth, that not a single living example exists to testify the creation of such

RESULTS OF NUMERICAL EQUALITY IN TRIBES.

a race. If, then, all the fossil *Pachydermata* were alive, and were incorporated, according to their affinities, with those now living, the contents of the whole group would probably be augmented to four or five times its present number; and those chasms, which now appear so wide, would be proportionably lessened; nay, it is highly probable they would not be greater, in proportion, than those between the different genera of the parrots. But, secondly, let us suppose that it was essential to the symmetry and harmony of nature, that all her groups of the same rank and value should contain pretty nearly the same number of species, and that their numerical contents should be proportionate to their value. What, in the present instance, would be the result? The tribe of *Scansores*, or climbing birds, includes the parrots; and, upon a rough estimate, certainly contains between four and five hundred species. We know, by induction, that this tribe is equivalent to that of the pachydermatous quadrupeds. Now, if these tribes were as equal in their contents, as they are in their rank, more than half the earth would be overrun with monsters. Elephants would be as common as flies; we should have to reckon not *two*, but perhaps two hundred species. All the large rivers would be almost choked with hippopotami. Rhinoceroses would swarm in the woods, in herds of thousands, as the parrots do now in the forests of America. And huge megatheri, perhaps of a hundred species, would attack a forest, and strip it of its verdure in a few days. The world, in fact, would be filled, as it once was, with monstrous animals; and man would find no resting place in it.

Nor is this all : the whole of these gigantic creatures feed upon herbage, grass, or the leaves of trees. Let us imagine, then, for a moment, what would be the state of those countries, as the vegetable world is now constructed, which should be inhabited by thousands of such monsters, as the tropical regions now are by the parrots. The consumption of food necessary to support such creatures would be enormous. No plains would be sufficiently fruitful to graze thousands of elephants and rhinoceroses of hundreds of species. The trees would be bared of their leaves, and verdure would disappear. The earth, in fact, would be as much devastated as if perpetual swarms of locusts had stripped it of its clothing ; and thousands of these devouring monsters would annually perish for want of food, poison the air, and create pestilence and famine. Such results, however frightful, are too obvious to be denied. The paucity, therefore, of pachydermatous quadrupeds, instead of proving a want of uniformity and consistency in the groups of nature, is the very peculiarity which manifests the harmony and design with which they were balanced and adjusted, by Infinite Wisdom, from the beginning. The pachydermatous quadrupeds, considering their immense size, are proportioned to the rest of the animal creation, throughout which we find that great bulk is restricted to few individual forms, while excessive minuteness is extended to countless millions. What, therefore, would at first seem to constitute the *Pachydermata* an imperfect group, is, in reality, its highest perfection. If its chasms were fewer, or narrower, it would possess more forms, for which the world, in its present state, could scarcely find room. Be this, however, as it may, we need not, after this, require further demonstrative evidence to prove the inequality of numbers in natural groups of the same value; or that apparent gaps may not often be accounted for on the soundest and most philosophic principles.

CHAP. V.

ON THE CHARACTERS OF NATURAL GROUPS.

(162.) THE characters by which natural groups, like those we have hitherto contemplated, are to be known and designated, has been a fruitful subject of disquisition among writers. It has been customary, until within the last few years, for naturalists to decide, *à priori*, upon those characters which a group of species, or a single one, should possess, in order to constitute a genus. This mode of proceeding, as may naturally be supposed, led every one to follow his own opinion; so that almost every part of an animal, in turn, had been singled out as the most important for this purpose. Thus, Linnæus founded his genera of birds entirely on the form of the bill and the construction of the feet; totally disregarding the formation of their wings,—which is one of the chief characteristics of birds,—and entirely overlooking their manners, habits, and food. In entomology, however, he constructed his genera on a totally different principle. Here he considers the wings of insects as affording the most important characters; and he has accordingly founded all his great divisions, and most of his lesser ones, in the different modifications which these members present; while the mouth and the feet, which were so highly regarded in his arrangement of birds, are scarcely noticed in his classification of insects. Fabricius, on the other hand, as if determined to fly to the other extreme, takes all his leading characters from those parts of insects which his great master regarded as insignificant. While some of the French naturalists, looking chiefly to the feet, built their systems on the number and form of the joints they contain. Whether an insect or a bird fed upon animal or vegetable food; whether it lived upon the ground, or habitually avoided it; or whether it flew with celerity or with difficulty; were matters which then had little or no influence in the determination of groups: indeed, they were almost thought too trivial to notice. True it is, that in very many instances, natural groups were still preserved; but, generally speaking, as there were no determinate principles for classification, so there could be no uniformity of arrangement, or consistency of separation. The most heterogeneous combinations, of course, resulted; of which the group of *Scarabæus*, as left by Linnæus, and as still exhibited in more modern works, affords a striking instance. Compare these crude and almost unintelligible arrangements of insects with the lucid, harmonious, and philosophical exposition of them given in the *Horæ Entomologicæ*, and the unprejudiced entomologist will at once see the difference between arrangements which lead to nothing, and arrangements which are in harmony with the primary laws of nature.

(163.) Now, the objections against different writers employing different characters for the same divisions, is not in the simple fact itself, but because they aim at no other object than to abridge the labour of research, by dividing and subdividing; and because, when that is done, we are left without any ulterior result or generalisation. Let us look to a case in point. A modern German entomologist, taking the old genus of *Curculio,* or snout-beetles, divides it into what he calls *genera,* amounting to about two hundred. Now these divisions, in a family so vast, may very probably facilitate our search after a particular insect; and so far may be very useful to the mere nomenclator. But the first questions which the philosophic entomologist will ask, are these;— Upon what general principles are these groups founded? and how far are the same principles applicable to other families? What are the results obtained by this new mode of arrangement? and how do they bear upon other approximating assemblages? If no general principles have been aimed at, or can be deduced, and the only result obtained is that we may more readily find the name of an insect, it is clear that the very first principles of true science have been lost sight of; and that if groups are to be so formed, natural history is but a study of words and names. Another writer, coming after, and choosing to draw his characters from a different set of organs, may divide this family into four hundred such genera; and, if we annex no definite meaning to the term, who can object to this? If a timely check is not given to this mania for making divisions, and calling them *genera,* we may very probably see the above supposition actually verified.

(164.) It is seen, by reference to all the best classifications, that scarcely two writers, even in the same department of zoology, agree in drawing their characters from the same organs, or from the same premises. And, indeed, if we consider the subject for a moment, it is impossible they should; because such divisions rest only upon individual opinion, without reference to any common standard by which such opinions can be judged. It is clear, also, that the same organs will have different degrees of consequence in different classes. By taking the form, number, and disposition of the teeth into consideration, we bring the quadrupeds into large but very natural divisions; but to carry on the same set of characters to fish, and make *their* dentation the chief guide in their arrangement, would be manifestly absurd, if not impossible. Hence it follows, that not only in a natural, but also in an artificial system, there are no organs in animals which can be universally employed to furnish generic characters, and to which we must exclusively direct our attention. The question, then, arises, by what rules are we to be guided in defining such divisions, and in giving them a stability which artificial groups have not?

(165.) When, therefore, the naturalist, following the principles already detailed, has before him a generic group, whose affinities, more or less, appear to be circular, he is next to seek for those characters which are most prevalent in all the forms or species which compose it. It is a matter of perfect indifference, what organ, or what set of organs, furnish these characters, provided they are more comprehensive than others, and are of such a nature as to be readily detected. His great object, in fact, is to point out, with clearness and precision, how the group before him is distinguished from all others; and if he can do this effectually, it matters not by what means the object is accomplished. He is not, however, to expect that he can so far isolate a natural group, as that there shall be no exceptions to the characters he assigns it; or that each of the individuals composing it shall possess those characters in the same degree. This would imply not only the existence of absolute divisions in nature,— which all experience is opposed to, — but would be directly at variance with what has been just said on the chain of continuity or gradation; for where there is gradation in structure, there must be gradation in character. It will be sufficient, if the greater portion accord with his definition, and if the others present a gradual diminution of the same set of characters, fading and blending into others belonging to adjoining groups. In searching, therefore, for such characters, we must take into the account every circumstance that is known regarding the economy and the structure of the objects themselves; and from all these make a selection of such as are most constant, universal, and obvious. It will almost always be found, that a peculiarity of internal organisation is accompanied by a corresponding difference in external structure, and that both these are adapted for that particular mode of life which the animal pursues. As there is a constant harmony between the conformation of an animal and its peculiar economy, we should study the former with a constant reference to the latter, but yet draw our characters from the *first* rather than from the second; because form can be always determined, whereas habits are only to be traced from the living subject. Every one, for instance, can see whether a bird has its claws acute and very much curved, or whether they are comparatively straight and obtuse: now we know, from observation, that these modifications indicate two very different habits; the first belonging to birds which always perch upon boughs, and the latter distinguishing such as live chiefly upon the ground. These habits, however, can never be known to the student as matters of fact, who merely sees such birds in a museum: we are therefore to direct his attention, in the first place, to circumstances or characters which it is in his power to see and verify; and afterwards to show the particular influence of such characters on habits and economy.

(166.) The essential or most prevalent characters of our group, *as a whole*, having been ascertained, we are then to examine it more in detail, tracing the mode in which these characters vary, and ascertaining how far, and in what way, this variation is accompanied by a difference of habit and economy. There is, for instance, a whole family of beetles (the *Petalocera Saprophaga* of M'Leay) which feed on *living* vegetables, in contradistinction to another, which devour them only in a *decayed* or putrescent state: but among those which agree in feeding upon living plants, we find some restrict their diet to the petals of flowers, others select only the green leaves, and many live upon the juices of the stem. Here, then, we have modifications of the same habit; and it is our business to trace such variations, whether in form or in economy, through every natural group. Now, as there must, in the natural system, be a harmony of design, that harmony, if it is not universal, and extending to the most minute particulars, cannot be worthy of those attributes belonging to the God of Nature. We must not, therefore, content ourselves with noting the variations we are speaking of, and viewing them as simply confined to the group in which they occur; for this would be taking a narrow and confined view of things. Our business is to trace them in all other groups — not only such as are adjoining, but in those belonging to the same great division of animals: we must, as it were, ascertain how far they are amplified and expanded ; and trace their prevalence in as wide a circle, and through as great a number of other assemblages, as possible. We should ever bear in mind, that nature every where presents those two kinds of relations already explained, namely, affinity and analogy; and that both these universally belong to *all* groups. Hence we may conclude that there must be a certain order in which analogies occur, and that the series in *one* group will occur in precisely the same order in another. Were it otherwise, there would be a want of harmony, utterly inconsistent with that ideal perfection which we attach to the system of nature. Accordingly it has been found, that in a number of ornithological groups, these analogies do actually occur in precisely the same order, and with the same regularity, as the seasons of one year follow and correspond to the seasons of another. It is, then, to these modifications of form which every circular genus presents, that we give the name of sub-genera (144.). Now, the characters by which such divisions are to be defined, are most judicious when they are drawn from two structures: first, from that structure of the animal itself, which it possesses in common with all others of the genus; and secondly, from that peculiarity which renders it analogous to many others out of its own genus.

(167.) An instance, taken from an ornithological group, will illustrate the foregoing remarks. The genus *Sylvicola* (*North. Zool.* ii. 205.), or titmice warblers, comprehends so large an assemblage of species, that we are able to trace, and demonstrate, its circular succession. All these birds agree, more or less, in the form of their wings; the first quill of which is nearly as long the second. This being the most prevalent character, we select it as the chief mark of discrimination, without stopping to enquire whether other groups are to be also characterised by these organs; for *this*, we see, most assuredly *is*. We find, however, that among all the birds thus brought together, we have different modifications of the other organs: some have the bill more conic and entire, others have it depressed and notched; in two or three, it is very sharp pointed, and even curved; while a few depart from all the rest, in having feet adapted for climbing. Now, it is manifest, that if we had set out with presupposing that all birds were to be classed by the bill and the feet, and that all other characters were of little or no moment, this most natural group would never have been detected. It is also clear, that, being discovered, we cannot draw our essential and primary characters from any of those organs which, as above stated, are seen to put on so many different forms. Yet, although we are unable to employ these variable peculiarities in a *primary* sense, they afford admirable distinctions of a *secondary* nature ; and these, when coupled with the peculiar formation of the wing common to the whole group, give us a compound of characters by which all titmice warblers may be distinguished, almost at a single glance, from the hundreds of species composing the family *Sylviadæ*.

(168.) Now if, in proceeding to the investigation of another genus, we find *that* also characterised, as a *whole,* by some one peculiarity of structure; and that it also comprehends subordinate forms, more or less agreeing with those in the last; we have every solid reason to suppose these subordinate forms, in both groups, to be analogous; or, in other words, to represent each other. To give them, therefore, discriminating characters, we unite that of the entire genus to that of the sub-genus, as before intimated; and these, collectively, give us a distinguishing formula, by adhering to which we cannot possibly err.

(169.) Of natural groups, Linnæus certainly had a very sound theoretical idea, when he said, that every genus would furnish its own characters, and not that the characters should form the genus; thereby implying, that we were first to place objects together which appeared closely related, and then to discern what were the peculiar and tangible characters which made them so. The truth is, that, generally speaking, an unscientific person, but with a discriminating eye, is much more likely to assort objects into natural assemblages, than one whose ideas are shackled by the dogmas of nomenclators, and the prejudices of systematists. Nature, in the midst of her astonishing, endless diversity of forms, still seems to delight in preserving a marked degree of uniformity and consistency in her own groups; not only in regard to their habits and general structure, but in such things as are most likely to strike common observers — such as size, colour, and geographic distribution. She rarely, if ever, places in the same genus, animals of any striking disproportion in their dimensions. We have, for instance, no eagles of the size of thrushes; nor any finches, out of some hundreds, that are larger than sparrows. The typical gallinaceous birds, as the peacock, pheasant, Turkey fowl, &c., are all large; and have so many points of general resemblance, that the ordinary observer, caring nothing for systems, sees at once that they all belong to a natural group. It matters not, in the first instance, whether we call such a group a genus or a family, because the rank it holds in the scale of creation is a subject for ulterior research: when, in reality, this rank is to be determined and demonstrated by an extensive analysis of all the other groups, large and small, in ornithology. Looking to the gallinaceous birds above named, we immediately perceive that, although they belong to the same family, they are of different genera: a peacock is as much unlike a pheasant, as a turkey differs from a fowl. We therefore proceed to define all these differences, making their distinctions from such characters as are most striking and most intelligible. The peacock is known by its immense fan-shaped tail; the pheasant, by having the same member long and pointed; and the turkey, again, is pre-eminent for its naked face, fleshy horn, and wattles. Here we perceive the force of the Linnæan axiom. We take a confessedly natural group like the present, and discover what are its general characters, and then descend to its variations. Had we done the reverse, and set out upon a theory that a fan-shaped tail, or a pointed one, or a naked face, were not to be admitted as generic characters in *any* group, we should be proceeding upon an arbitrary opinion, the absurdity of which would be manifested in the case before us; because, by acting upon it here, we should be obliged to distinguish a peacock from a turkey by some obscure and inconspicuous characters, which none but the comparative anatomist or the professed ornithologist would understand. Such a system might, indeed, be intelligible to them, but it would be almost useless to the great bulk of mankind.

(170.) The best characters for groups are those drawn from their external aspect; and it matters not in what this peculiarity of aspect consists, because it is almost universally accompanied by minute points of difference, which, upon strict examination, are sure to be detected. The latter, however, should not be brought in the foreground, and placed before the former, merely because it has been the custom for systematists to attach a fancied impoitance to minute characters, and to neglect those which will answer the same purpose of distinction, and yet be obvious to every one. If, for instance, the saprophagous and thalerophagous beetles can be equally well distinguished from each other by external characters, which require no dissection of their mouths; such a mode of discriminating them is to be preferred before all others, for the best of reasons — as being the most simple and obvious. Nor should we be tempted to employ anatomical characters, or such as are taken from the different modifications in the masticating organs, until we are absolutely compelled to do so by the failure of other resources. This, indeed, is in direct contradiction to the usual mode of proceeding pursued by modern naturalists; but, in the present state of natural history, and, indeed, of all science, it appears to us that one of the chief objects of its professors should be as much as possible to simplify. The science they would teach, and which they of course desire that others should learn, can only be rendered inviting to mankind in general, by being divested of all verbose technicality and minute investigation, not absolutely essential. If the same object can be arrived at by two roads — one smooth and comparatively easy, the other intricate, winding, and difficult — no one, in his rational senses, would choose the last in preference to the first. The same analogy should be pursued in science. Simplicity, perspicuity, and brevity should be the characteristics of all systematic distinctions, whether of groups or of species; and the more we study nature, the more shall we find that in this, as in all other branches of physical science, the laws which are most simple are at the same time the most universal.

(171.) Essential characters, or such as pre-eminently distinguish a group from all others, are generally few, and are usually confined to two or, at the most, three particulars; sometimes, indeed, where the group is much diversified, to only one. Thus the *Sylvicolæ*, already alluded to (167.), may be distinguished among the warblers solely by the form of their wings: but, if we wish to define them more decidedly, and to detach them from *all* other birds, we must, in addition to their own peculiar character, add that of the family to which they belong, namely, the warblers: we thus get the union already spoken of. By one we separate the *Sylvicolæ* from all other birds excepting the warblers; and by the other we point out those peculiarities which make them a particular division of warblers. This mode of definition is equally applicable to every group in nature, from the highest to the lowest. Where we can meet with three strongly marked characters, they may safely be employed; but one or, perhaps, two of them will always be found less universal than the other. When we come to the confines of a group so distinguished, the characters laid down for it gradually disappear, until at length only one out of three will be detected; that, therefore, which is most universal, is the most essential.

(172.) By simplicity of definition is meant, not a mere form of words, however desirable that may be, but the employment of such characters *only* as are necessary to the determination of the group, or object, in question. Thus, the family of *hornbills* or of *toucans*,— the one known by protuberances on the bills, the other by the excessive size and smoothness of theirs,— are sufficiently detached from all other birds, simply by these circumstances. We need not, therefore, in giving essential characters, go on to describe other points of structure; because they are not only unnecessary, but they distract the attention from those circumstances upon which it should be entirely fixed. So, in like manner, the genus *Sylvicola* needs only to be characterised by the bill and the wings; all its other characters being common to the next group in rank, of which it forms a part. In the lepidopterous order of insects, the form of the wings will in almost all cases determine the sub-genus; although in a monograph, or complete account of the insect, every one of its characters should of course be described.

(173.) To attain this simplicity, however, is much more difficult than would be at first imagined. For, as we find that no one peculiar set of organs can be universally employed for such distinctions; so it becomes necessary to discover, in the multiplicity of characters which every group presents, what is that one which is its peculiar and exclusive distinction. Now this, as we before remarked, can only result from extensive analysis, or by generalising the mode in which natural groups are seen to vary. It is remarkable, in every natural group of the diurnal *Lepidoptera*, whether the group be large or small, that there is one modification in which the lower wings are always more or less tailed. Numerous and very striking instances of this have been given in the *Zoological Illustrations*, selected from families and genera so different in themselves, as completely to do away with the idea that these swallow-tailed butterflies have any real *affinity* to each other, however strongly they are related by *analogy*. Now, having ascertained such a fact as this, through an indefinite number of groups, our next business is to see how far it can be traced in groups of larger dimensions; and finally, whether it is prevalent in quadrupeds, birds, and other vertebrated animals. Should we succeed in this, we obviously demonstrate that, through all her variations, nature has preserved, at least in one instance, a definite plan of variation, consisting in this, that in every natural group she gives to one of its types a preponderance of tail, or caudal appendages representing a tail. We maintain not, here, that such is actually the fact; we are merely stating a case, to illustrate the mode of *generalising the variation of characters* just recommended, and thereby simplifying the diagnosis by which the different forms in nature are to be distinguished. That this species of generalisation is not impracticable, at least in ornithology, has been clearly domonstrated in numerous groups defined on these principles in the *Northern Zoology*. And it follows, as a matter of induction, that if the variations of one extensive class of animals are regulated by certain general laws, manifest in all the groups of that class, the same will be discovered in other portions of nature, so soon as they have been investigated with sufficient attention to such circumstances.

(174.) Yet, although no general rules will here be laid down for the discovery or selection of *essential* characters, experience has shown that they may be derived with more chance of success from some circumstances than from others. It becomes desirable, therefore, to state what these circumstances are, and to trace the influence they possess in guiding us to sound and logical deductions. They relate more especially to the form or general contour, to the organs by which food is taken, and to those of locomotion. We will now give to each of these a separate consideration.

(175.) *The general form* or contour of an animal is that circumstance which first strikes the beholder, and impresses him with its peculiar character as a distinct being. Upon this, therefore, we have already laid great stress; and the more we become acquainted with natural groups, the more shall we be impressed with the importance of making *this* one of their primary distinctions. Thus, we see that thick and heavy animals are never naturally associated with such as are slim and agile. The typical ruminating quadrupeds, for instance, are large and heavy; and to expect that any of the species of oxen should have the light and elegant form of the horse, would be as inconsistent with the order of nature, as to see a mouse slowly and heavily pacing about our kitchen. Again, vultures, among birds, are nearly all characterised by heaviness of body and slowness of motion; whereas the whole family of hawks are proverbially quick and daring. Among insects, likewise, we see similar habits accompanying similar forms. What beetles are more slow than the *Meloe*, the genuine *Chrysomelæ*, and the *Geotrupidæ*, — the last, better known to our readers as the " shard-borne beetle with his drowsy hum," immortalised by Shakspeare. In opposition to these, we have the greatest developement of agility and grace shown in the shining *Cicindelidæ*; and in those glossy little *Carabidæ* which cross our path in a summer's day. Bulk, therefore, is generally connected with peculiarity of motion ; and both are highly characteristic of natural groups or types.

(176.) The unusual developement of any particular part of the body, unconnected with those we shall hereafter touch upon, comes under the general head of form or contour, and will be found of much importance in definitions. We never find, for instance, that animals, whose muzzle or face is greatly prolonged, are naturally grouped with such as have these parts short and very obtuse. Among quadrupeds, there are many striking instances of this law of nature. The muzzle of all the ant-eaters is so excessively lengthened, that it seems pulled out, as it were, into the shape of a rostrum or beak, such as we see among the curlews : we trace this peculiarity, again, through the whole family of shrew mice, and in the moles, and hedgehogs ; and, as if nature resolved that this type should not be lost even in the ungulated order, she preserves it clearly in the common pig. Among birds, we trace the same analogy of structure under a different modification. The muzzle of birds is, in fact, their bill ; and the excessive length of this part is one of the chief characters of the whole order of *Grallatores*, or waders. Look to all the types of this order, scattered in the rest of the feathered creation, and we find there are always some which have a more curved and lengthened bill than any of their companions : but the analogy does not rest here : great elongation of muzzle is always accompanied (for what reason has not yet been explained) with small eyes ; and these are placed very far back on the head. The most striking instance of this is seen in the elephant; but it is no less remarkable in the ant-eaters, the shrews, and the mice; while the common term of " pig-eyed," proverbially applied to small-eyed people*, expresses the singular fact we are now illustrating. The whole family of humming birds have the longest bills, and the smallest eyes, in proportion to their size, of any birds yet discovered; and these latter organs are placed so far from the naked part of the bill, that they seem central between the nostrils and the nape. When, therefore, any such particular structure can be traced through animals of different genera, and even in totally different classes, we feel assured that it must exercise an important influence in the system of nature, and that especial regard must be paid to such circumstances, in characterising natural groups. Similar results would attend an investigation of all such animals as have, in opposition to the last, very short and thick muzzles. Appendages to the head, whether in the shape of horns, crests, protuberances, or wattles, may be considered as furnishing strong and highly valuable characters of primary import. Horns are one of the chief characteristics of the ruminating order of quadrupeds — *Ruminantia* — in which is comprised not only the family of oxen, but the deer and antelope races. In these, the horned structure is at its maximum of developement, and consequently assumes the rank of an essential character. In the *Pachydermatæ*, or thick-skinned quadrupeds, belonging to the same order, we have a

---

* It is very singular that "pig-eyed" people have very generally a *long*, but never an *aquiline* nose.

modification of the same structure exemplified in all the species of rhinoceros; while the tusks of the elephant are no more than teeth, performing the same office, and applied to the same uses, as the horns of the ox. Now it is very remarkable, that naturalists agree in placing all these quadrupeds close to each other; or rather, in one of the primary divisions of the class: so that, with the exception of the morse and the monodon, or narwhal, the whole of the horned quadrupeds are found belonging to one natural order. This circumstance, of itself, is a strong corroboration of the opinion here expressed, and should lead us to infer that horns and such like appendages indicate one of the essential characters of such groups or forms as possess them. But where, the student may exclaim, are we to look for horns among birds? for, if such appendages really constitute essential characters, they must either be found in other vertebrated animals, or a structure, limited to one class, can never be ranked as one of the primary types of nature. Now, the only family of birds which may be said to possess horns analogous to what we see among quadrupeds, are the hornbills, *Buceridæ*, nearly all of which have excrescences, as they appear, rising from the front of the bill; and one of the species is so remarkable in this respect, that it is called the rhinoceros hornbill. Other birds, — as the Tragopan pheasant of India, the horned screamer of America, and the unicorn chatterer of Brazil, — have hornlike protuberances, but they are soft and fleshy. The truth, however, appears to be, that horns are represented in the feathered tribes by crests, which are not merely ornaments to these elegant creatures, but are actually used by them to scare and frighten away their enemies. On this curious fact, hitherto unrecorded, we shall subsequently enlarge; yet this analogy, being established, shows that the crested and horned structures are synonymous; and that, under one modification or the other, it is as prevalent among particular groups of birds, as of quadrupeds. On turning to insects, it is no less conspicuous: here the horns assume a very decided character; and although given to numerous insects, scattered in all the families, are more especially developed in the gigantic beetles forming the modern group of *Dynastidæ*. In the soft-winged flying orders, especially in the *Lepidoptera* (more analogous to birds than is any other), crests take the place of horns, of which the whole tribe of the *Noctuidæ* is a striking example. When, therefore, in a group of animals, we see that horns or crests distinguish the major part, we draw their essential character from that circumstance: but when, in another group, these appendages are only confined so a small portion, we take the essential character from the general peculiarity of the whole, and discriminate these few which are horned as a subordinate assemblage.

(177.) Another of the most prevalent forms in the animal world is that in which the tail, or the caudal appendage representing it, is excessively developed. In looking, however, to this member, we must carefully note the peculiar sort of developement it presents; because, although it may be very large in any given number of examples, its formation, and consequently its offices, will be totally different in different individuals. Thus, in the horse, where the tail is more developed than in any other quadruped, it may be looked on as more ornamental than essential : at least, we know it is not connected so intimately with the habits of the animal, that its loss leads to the impossibility of supporting life. But among the long-tailed monkeys of America, this result would inevitably follow ; their tail is prehensile, and, by being employed as a fifth hand, in climbing trees, is absolutely essential to them in procuring the fruits upon which they live. We have examples precisely analogous to these, in the class of birds. The glory of the peacock is its tail; it is, indeed, a splendid ornament, but it is an ornament alone. And we know, from the habits of these birds, not only in a state of captivity, but in their native regions, that they can search upon the ground for their usual food just as well without their tail as with it : but with the parrots, and more especially the woodpeckers, the case is different; here, also, the tail is highly developed, but in a very different way. In the parrots, it performs the double office of aiding flight, and the power of climbing. Those who have witnessed the lofty and arrowy course of the splendid mackaws of the New World, know that this celerity of motion would be utterly lost if the tail were of any other structure ; while every one who has watched a parakeet, even in the confinement of a cage, may have remarked how much this member contributes to facilitate the habit of climbing possessed by these elegant creatures. Still, however, the utility is only seen at its maximum in the most typical climbers, or the family of woodpeckers. Here nature throws aside ornament, and makes the tail of these birds not only useful, but absolutely essential to their means of supporting existence; the loss of it, to a woodpecker, would, in fact, lead to the loss of life. The bird could no longer climb trees in search of food, because it would want that support in a perpendicular position which the tail supplied; so that, like an American monkey so circumstanced, it would die within sight of ample nourishment. There is still another form under which a great developement of tail is observed, and the use of which is ·exclusively confined to the flight. No instances of this form are found among quadrupeds, but there is scarcely any family of birds that is without it. In the modifications just described, the shape of the tail is always round or wedge-shaped, but in that we are now speaking of, it is invariably forked. A familiar example of this is seen in the swallow family, where it is most prevalent, but nowhere is it carried to such an extent as in some of the goatsuckers of tropical America, and the fork-tailed kite of the United States. It does not appear, however, that this structure is so prevalent in all the individuals of a natural genus as those already noticed; for many of the swallows and most of the goatsuckers have even tails: hence this character, among birds, can rarely be employed otherwise than to designate sections, or perhaps sub-genera: in such cases, however, it becomes essential, or the chief mark by which such forms are to be pointed out.

(178.) Among the winged insects (*Ptilota* Arist.), great length of the posterior wings, or caudal appendages to the body, always indicate a sub-genus, and will sometimes point out groups of a higher denomination. It is remarkable that the lower wings of the *Lepidoptera*, when thus unusually lengthened, perform the same office in flight as the tail does among birds, for we find that all the swiftest flying butterflies have what are aptly and justly called swallow-tailed wings; that is to say, their extremities are lengthened out into tail-like processes. The sub-genera *Podalirius, Protesilaus, Leilus,* and *Eudamus* (all of which are figured and characterised in the *Zool. Illustr.* 2d series), are striking examples of this form, and the *Eudamus borealis* (*Hesp. proteus* Lin.) is such a common species that almost every cabinet contains an example. Essential characters, therefore, may safely be drawn from this structure, for its universality among the classes of insects and of birds leads us to infer it is one of those primary types of variation which nature has herself chosen. On looking to the other orders of winged insects, we find but few examples of elongated lower wings, and these are chiefly confined to the *Neuroptera* Lin.: but here we find that caudal appendages are almost universal, so that they nearly become one of the essential and *natural* characters of the whole order. The entomologist will observe that we are now speaking of the *Neuroptera*, as defined by Linnæus; and not of that section of it only to which modern systematists, on views the most artificial, have restricted the name. It will be hereafter shown that this order, as contemplated by Linnæus, is one of the most natural in the whole circle of the *Annulosæ;* and that it never would have been dismembered, had the analogies of the group been sufficiently studied. The caudal appendages, then, of the *Neuroptera* assume different forms, suited to their different offices: in the *Forficula,* or common earwig, they are used as means of defence; in the locusts, they are employed to perforate the ground for the deposition of the eggs; and in the dragon-flies (*Libellulidæ*), they are connected with the process of impregnation. In the ephemera, they assume the appearance of three long, hair-like bristles, and really become tails, analogous, in appearance, to those of the ichneumon flies; but their particular use seems not clearly understood. Sufficient, however, has now been stated, to show the importance of using such characters in a generic sense; and wherever they occur in other groups, there will be no danger in employing them as the essential indication of sub-genera.

(179.) Let us now consider the value of characters founded upon the structure of the mouth. These, it is obvious, are, when rightly used, and constantly viewed in reference to the nature of the food, of the highest importance. Their value, however, entirely depends upon these considerations: if, for instance, we were to set out with placing all birds that live solely upon fruits in one division, all those which fed upon insects in another, the seed-eaters in a third, and so on, we should have an arrangement perfectly unintelligible. Again, if all flesh-eating birds were to be separated from such as eat fruits, we must exclude several of the American buzzards, which, as it is asserted, feed as much upon one as upon the other. Such deviations from the general character of rapacious birds may be used as a subordinate mark of discrimination, if accompanied by deviations in structure, yet not otherwise. The nature of the ordinary food of an animal is almost always indicated by its external organisation. We know by experience that certain habits of life are indicated by certain peculiarities of form, so that by studying the conformation of an animal which we have never seen alive, we can arrive, with a degree of certainty almost incredible, at a general knowledge of its habits and economy. This is more particularly the case among birds; because, perhaps, they have been more especially studied with reference to these circumstances than any other class of animals. In these the form of the bill is of as much importance in determining the food, as are the teeth of quadrupeds or the jaws of insects. Purely insectivorous birds must always be considered as of a different type to such as partake more or less of vegetable food, even should such deviations be found in the same subfamily or even genus. A singular instance of this occurs in the sub-family of the Titmice (*Parianæ*). All these birds live entirely upon insects excepting one genus, that of *Accentor*, to which belongs our common hedge-sparrow: this bird, as every one knows, feeds as much upon small seeds as upon minute insects; yet it is so intimately and unquestionably related to the group it has been placed in, that the perfection and unity of the whole division would be destroyed, if, because it was not purely insectivorous, we took it out of the circle, and endeavoured to find a place for it elsewhere: nevertheless, we still make use of this peculiarity of habit

STRUCTURE OF THE MOUTH.

as an essential character of *Accentor*, because it is manifested by an external conformation, which indicates such habits. This instance, out of numberless others, is a convincing proof that not even a difference in the nature of their food will invariably or *completely* detach insectivorous from granivorous birds. On looking, however, to the great divisions in every class of zoology, we see that no characters can be more natural than such as separate destroying from harmless animals among quadrupeds. The most ferocious genera are brought together in the order *Feræ*, composed entirely of the beasts of prey: here we have the lion, the tiger, and all the races of leopards, panthers, and cats; together with the weasels, polecats, and those minor blood-sucking quadrupeds, as destructive and sanguinary towards the smaller animals, as the former are to the larger. These find their representatives in the rapacious order of birds; and in both, the nature of their food is at once explained by the construction of their mouth: the teeth of one, and the notched bill of the other, being especially adapted for tearing flesh. Extending this analogy to the insect world (which, by the way, has never yet been done correctly), we find the great majority of *Aptera*, or the wingless orders, — as the spiders, scorpions, crabs, &c. — feeding in like manner upon other insects, and living only upon the blood and flesh of their victims. So strongly, indeed, has nature preserved this distinction between her types of *evil* and of *good*, or, to drop metaphor, between noxious and innoxious animals, that not only are these the essential distinctions of her primary groups in every class, but they can be traced downwards, into all the different orders which compose a class. In proof of this we cite the *Hemiptera* as the rapacious order of the *Ptilota*, the *Libellulinæ* as the analogous group in the *Neuroptera*, and the *Prædatores* (*Chilopodomorpha* M'Leay) as the corresponding representation of the *Feræ* in the order of *Coleoptera*. We merely intimate these novelties in natural arrangement (which will be separately treated of hereafter), to show the importance of essential characters founded upon the food of animals, or rather on the structure by which its nature is indicated.

(180.) Animals which are omnivorous, or devourers both of animal and vegetable matter, present a singular union of those characters which respectively belong to the groups just mentioned. Unless, therefore, we have a previous knowledge of the circumstance, it becomes extremely difficult to determine, from the simple examination of a dried specimen, what was its natural food. The crow family (*Corvidæ*), which are the most perfectly organised of all birds, are of this description. They feed upon almost any thing which has life, either animal or vegetable, and even upon carrion. The toucans and the trogons belong to two different tribes; but as they are the points of union connecting the *Scansores* and the *Fissirostres*, we find that both partake of the same vegetable and animal diet. The family of rats are likewise omnivorous, and by this peculiarity they make the transition easy from the carnivorous to the herbivorous quadrupeds. Such of the tyrant fly-catchers of America as show a decided affinity to the true shrikes, feed, like them, not only upon insects, but lizards and other reptiles; and Azara asserts that the *Saurophagus sulphuratus* or Bentevi fly-catcher of Brazil, actually picks the meat from the bones of such carcasses as have been left by the larger animals of prey. Such facts are highly interesting, and will frequently, as in the bird last mentioned, decide a doubtful point in natural arrangement: it follows, therefore, that omnivorous habits furnish characters of great value.

(181.) There is another mode of taking food, very general among the invertebrated animals, but not so distinctly marked in the higher classes. We allude to suction, by which fluids alone constitute the sustenance of the animal. There is a modification of this structure of mouth in the ant-eaters, and in the honey-sucking birds (*Tenuirostres*), where the tongue alone is employed to collect food; but the most perfect examples of the suctorial structure of mouth are found among the four-winged insects, where we find two entire orders, — the *Lepidoptera* and the *Hemiptera*, — entirely destitute of jaws, and deriving their sole nourishment from juices, sucked up either by a slender jointed trunk, or a long and spirally rolled proboscis. Nature has evidently made this structure a leading distinction of particular groups throughout the animal circle, for we find that in every class some are suctorial, while others are not; and that this habit is always accompanied by a uniformity in the general shape of the rostrum or mouth, which, as suited to such functions, is always very long and slender. This we see in the ant-eaters, the humming-birds, the sun-birds, the whole of the typical waders (*Grallatores*), the *Lepidoptera* and *Hemiptera*, the zoophagous *Testacea*, the suctorial *Radiata*, the worms and the leeches. These examples, it will be perceived, are taken almost at random from different classes of the animal kingdom; and clearly show that essential characters, founded on this particular structure, are of primary consequence.

(182.) We are now to consider the value of distinctions derived from the organs of locomotion, that is, from the feet and wings: these two members being represented in fish and other aquatic animals by *fins*. Each of these is entitled to a separate consideration. The most perfect developement of foot is found in quadrupeds, and the different modifications of structure which it presents are truly surprising: the feet of some are barely sufficient to enable the animal to crawl slowly and irregularly upon the ground; and even this, in the sloths, is obviously accompanied with pain. Some of the Lemurs, also, are equally incapacitated from the ordinary motion of quadrupeds. Yet, place these animals on trees, and they appear to be in their proper element — active, expert, and indefatigable; " they live, and move, and have their being," not by walking but by climbing. But the most accomplished scalers of trees are the monkeys, whose limbs, in fact, are more perfectly formed for this purpose than those of any quadruped in creation: the agility which these animals display in their native forests is really astonishing, and far exceeds that which they still retain in confinement. What a con-

ORGANS OF LOCOMOTION. 265

trast to these antics is the "measured step and slow" of the hoofed quadrupeds to whom the faculty of climbing is totally denied! Yet here again we have numerous modifications of pace, from the slow and stately walk of the elephant and the ox, to the fiery impetuosity of the horse, and the bounding spring of the elegant antelope. Yet these are not the most extraordinary of nature's contrasts. The continent of Australia presents us with two other modes of progression, totally different from those just mentioned, and almost confined to the quadrupeds of the southern hemisphere. These belong to the phalangers and the kangaroos: the first have their representatives in the New World, in the squirrels. The phalangers, in fact, are flying quadrupeds; not, indeed, from possessing wings, but from having their feet so united by a thin expansive skin, that they can take prodigious leaps from tree to tree, so as to give them the appearance, to an ordinary observer, of flying. This winglike membrane, in short, acts and folds up like a parachute or umbrella, and supports the animal in the air when it otherwise would fall upon the ground. In the kangaroos, on the other hand, the power of leaping is developed in a most remarkable manner: their fore-feet are so short as to be perfectly useless in running, but their hinder are enormous; and with these, assisted by their thick and powerful tail, they proceed by a succession of amazing bounds or rather leaps, repeated so fast, and so wide, that these animals, with two feet only, will generally escape from any other, even a horse, that has four. Two other modifications of foot remain to be mentioned: the first belongs to purely aquatic animals, like the otter, the seal, and the ornithorhynchus; to these the power of walking is almost denied, their feet being remarkably short, the toes connected, and the whole structure adapted almost exclusively for swimming: the second and last structure is restricted to the beasts of prey forming the order *Feræ;* the only material difference between these and the ordinary feet of five-toed quadrupeds, consists in the power they possess of retracting and protruding their claws at pleasure. The importance of this property to the animal is sufficiently obvious, when we consider the especial use which is made of them. Strong, and peculiarly sharp, they are employed as formidable weapons of offence and of defence; with these they seize their prey, tear it into pieces, and defend themselves from their enemies. We cannot have a better or a more familiar example of these habits, than in the domestic cat.

(183.) The feet of birds are no less varied, and afford us the means of discriminating the primary divisions of the feathered creation, without having recourse to any other help. Nor do their differences terminate here: under each of the five great modications, or types of formation, which may be seen in birds, are contained others, still preserving the essential character common to all, but deviating into minor types, which again point out little assemblages of groups, or of species, more especially united among themselves. The most perfect birds in creation are of course such as have the greatest complexity of structure, and the greatest variety of powers. These are unquestionably the perchers;

or such as have their feet adapted not only for perching on trees, but for walking upon the ground. Some unite both these powers in the greatest perfection. The crow, the thrush, the robin, sparrow, and numerous other familiar birds, are as often seen in one situation as the other, and in both are equally at home. But the swallow is rarely seen on the ground, save when employed, by the side of a wet puddle, in picking up particles of mud for its nest. The flycatcher also (taking our common grey species as a genuine example of the family) very rarely sets its feet upon the earth, and then but for a moment. The larks, on the other hand, as rarely perch upon trees; the ground is more peculiarly their element, and the wagtails do the same. Nevertheless, all these families come into the general order of perching birds (*Insessores*), because they have those external characters which so distinguish them, yet modified in different degrees and proportions. Nature, ever watchful over her creatures, always makes up to them in one way what she takes from them in another. Of what use or advantage would it be to the wagtail, that it should run up trees like a woodpecker, or fly with the swiftness of a swallow, when its natural food is placed close to the ground? If all birds were equally endowed, if all could walk, climb, run, fly, or swim with the same ease and with the same perfection, they must all have a similarity of structure adapted for such powers; and that variety, which is one of the greatest wonders of the creation, would never have existed. We therefore see gradations of the same quality—those, for instance, of perching or of walking, throughout nature; and these gradations are so numerous and so combined with other qualities, that the variety thus produced becomes infinite. Thus it is among perching birds: all have the hind toe (which corresponds to the thumb of the quadrumanous *Mammalia* or monkeys) placed on the same plane as the sole of the foot, by which means they are enabled to grasp an object from *behind*, with as firm a hold as they do *before* with the anterior toes: by this means the grasp is rendered firm; whereas, if only the fore-toes were employed, there would be wanted a support, or more properly a counteracting force, on the other side, to preserve the body steady. How important this structure is to the perching order, may be judged of by any one who should endeavour to grasp a broom or other round stick in his hand, by his four fingers *only:* he will think, perhaps, that it can be done very effectually; but a boy, with half his strength, will find no difficulty in wresting the stick from him. But if he again takes it, and applies his thumb in addition to his fingers, he will immediately perceive with what additional strength he now grasps the stick; and that no one, not physically stronger than himself, can take it from him. Applying this to the birds in question, we see that the toe of perching birds, like the thumb of the human hand, is on the same plane with the claws; and that both are more especially adapted for grasping *round* objects. This is why all weapons, or handles of utensils, are, for their more ready and convenient use, made *round;* and as the perching birds chiefly frequent and roost among branches, which are also round, their toes are especially adapted for grasping such objects. The feet of some, indeed, show an utter incapacity for *walking*, and probably even for *standing* upon the ground; such a structure is seen in the kingfishers, the bee-eaters, the trogons, and even the puff-birds, whose feet have a very peculiar structure: the anterior, or fore toes, are united together for nearly half their length, so as to form a greater breadth of surface on their soles, by which means, although they are deprived of all power to *walk*, they are better able to support themselves, as they do for hours, sitting almost motionless on a dry twig, watching for insects.

(184.) The most striking modification, however, of the perching structure of foot, is seen in the climbing birds; whose habits require that they should be possessed of a much firmer grasp than usual. Now this has been effected by the toes being placed in pairs, two forward and two backward, so that the counteraction of force becomes perfectly equal. The parrots, the woodpeckers, the toucans, and the cuckoos, are all distinguished by this sort of foot; which, while it enables them to climb with greater facility than any other birds, proportionally disqualifies them, as a necessary consequence, from walking with readiness upon the ground. How nimbly and how gracefully will a tame parrot, for instance, ascend and descend the wires and perches of his cage; yet open the door, and place him upon the ground, or a flat surface, and he becomes one of the most awkward and clumsy of all birds: he waddles, rather than walks, and appears as much out of his natural element as is a swan upon dry land. True it is that *all* parrots are not so formed; or rather there are some which, still possessing the same character of feet as those just mentioned, have them so modified and altered, that they can not only walk with perfect ease, but habitually frequent the ground in preference to other situations. It is curious and instructive to see how nature has effected the difficult object of giving to a scansorial foot the facilities of a walking one, without impairing the essential character of the family to which these ground parrakeets belong. On looking to the feet of the generality of parrots, it will be observed that the claws are particularly strong, broad, and well curved; so that when the foot is placed upon the ground, the tips of the claws touch, or come in contact with, its surface. Now much walking would soon wear away their points, as we see it does in those of gallinaceous birds; and this, to birds which climb, and use their claws for that purpose, would be a serious injury. But on examining the foot of a ground parrakeet (*Pezoporus formosus* Ill.), we see, indeed, the same general disposition of the toes; that is, two before, and two behind; but the legs are higher, more slender, and therefore better adapted for walking: the claws, moreover, are formed upon quite a different principle; instead of being thick and hooked, like talons, they are long, slender, and very slightly curved, so that when the bird walks, or (as we suspect) *runs*, on the ground, the points do not come into contact with the surface, which they unquestionably would do were their curve greater than it really is. From this peculiar conformation results two essential advantages to the bird : first, by having longer and more slender feet, it walks upon the earth with greater facility than any of its family; and, secondly, it does this without any pain, impediment from, or risk of injury to, the ends of its claws. This explanation, which has never been, we believe, before attempted, is one out of a thousand proofs of those gradations in the animal world which demonstrate natural series, but which some writers have had the hardihood to deny, as if the most acknowledged truths and the most obvious facts were to be made matters of doubt and of difficulty. We have before us, at this moment, a beautiful series of species, showing every possible link of gradation, from the ground parrakeets of Australia to those of tropical America, which we know, from personal experience, live wholly among trees. This, also, will serve as an example of that minuteness of investigation which the student is to pursue, if he wishes to draw just inferences from the structure in animals of whose habits and economy, when alive, he is entirely ignorant.

(185.) There are several other modifications of foot among the perching birds, which can here receive only a slight notice, sufficient to show the value of essential characters drawn from these organs. The structure of the claws, as just explained in the instance of the ground parrots, is almost sufficient in itself, for the discrimination of natural groups, or analogical types. Great curvature of these members indicates one of two habits, which are readily determined by other considerations. Either such birds are rapacious, in which case the claws are retractable, and are employed to seize their prey; or they are arboreal, that is, living for the most part among trees. Where the curvature, on the contrary, is but slight, it is a sure and certain indication that such birds chiefly live upon the ground. This may be considered a general law of nature; and perhaps a more familiar example cannot be cited than that of the rook, which seeks its food on the ground, when opposed to the jay, whose arboreal habits lead it to live and feed among trees. It will be perceived that the hind claw of the crow, when the foot is placed upon a level surface, is so much raised, that the tip or end is perfectly free; so that the bird can not only walk without unintentionally scratching up in his progress any loose stones or earth, but the ends of his claws, not coming into perpetual contact with other substances, are preserved sharp and uninjured. Let us now look to the foot of the jay: the hind claw is at once seen to be much more curved, and its point, when the toe is on a level, *touches* that level, and would obviously be injured by such constant friction, besides impeding the free walk of the bird whenever it moved upon the ground. By this simple character, therefore, we arrive at a knowledge of the different habits of these two birds; and by verifying this induction through all the other groups of ornithology, we get a general law of structure, which may be employed also as an essential character for any rank of groups. On the different modifications observed among the feet of the rasorial, wading, and swimming orders, we cannot here dilate: it will be sufficient to state, that, like those just noticed in the perching

(186.) Among insects (*Annulosa*), the organs of locomotion are varied in a surprising manner, according to the economy of each: our present remarks, however, will be confined to the feet. Celerity of motion, both in the air, and upon the ground, is rarely united; but, in the generality of the winged orders, is meted out in different and unequal proportions. The lepidopterous insects, which are, in truth, the typical perfection of the *Annulosa*, are the most perfect in their flight, although they scarcely ever walk. The *Coleoptera*, on the other hand, are the most active walkers, yet the most imperfect flyers. The *Cicindelidæ* and some of the *Carabidæ* show us a greater union of both these modes of progression than is to be found in almost any other insects. Their predacious habits, which oblige them to be constantly hunting and running down other insects upon which they feed, require this activity; and this is more perfectly accomplished, in the *Cicindelidæ* especially, by unusual powers of flight. Looking to these groups, and to the whole of the raptoreal tribe of beetles, we consequently see a perfection of structure in the foot suited to such a life: and this, with their carnivorous habits, is their strongest and best distinction. The five perfect joints in the legs of these beetles correspond to the five perfect toes of the insessoreal birds, and are an additional proof that a great developement of the foot is a sure indication of typical pre-eminence. Among the apterous class, we see this again in the numerous feet of the scorpion, centipede, &c. Progression, however, among insects is not confined to flying, running, or swimming; for although we lose sight of the leaping structure after quitting quadrupeds, we see this form again developed in the highest perfection both among the apterous (*Aptera* L.) and winged classes (*Ptilota*); of which the fleas in one, and the grasshoppers in the other, are examples occurring to every one. We cannot doubt, indeed, but that the saltatorial structure is one of the primary types or models of nature, for she has produced it, under an infinite variety of forms and modifications, in almost every group of animals. Of this description is the hopping of some birds, in opposition to those which walk; the hop being, in fact, but a short leap. Even among the *Lepidoptera*, where we should not expect to find any such analogous form, there are the *Hesperidæ*, which, from the sudden and peculiar quickness on the wing, have acquired the common but expressive name of *Skippers*. Not to mention the monilicorn tribe of beetles (under which we place the *Chrysomelidæ* of Linnæus), where we have hundreds of little species, familiarly known to our farmers as *fleas;* meaning thereby, a little black beetle of the same size, colour, and leaping in a similar manner to real fleas. A knowledge of all these habits, and of many others not alluded to, may be gained by induction, with almost as much certainty as if they were learned by experience. These inductions produce higher conclusions; and if these are confirmed by every thing we yet know, we arrive at a law of nature. That the saltatorial structure is one of those primary forms upon which all the variations of the animal world are modelled, will be hereafter shown more at large. Yet, as, on the plan we are now pursuing, the naturalist is to keep his mind free from the influence of every theory, we only wish to enforce upon him, from the above examples, the necessity of making all such deviations from the ordinary structure of the feet, a ground for separation and distinction, even if no other exists, between two forms in other respects perfectly similar.

(187.) Hitherto we have contemplated those animals only, which, with the exception of fish and of serpents, are provided with articulated feet; but as we descend to the more imperfectly or less organised groups, as the *Testacea*, the *Radiata*, and all those " slimy things" which inhabit the depths of ocean, no such organs exist, and locomotion is effected by other means, and in various ways. Some of these animals crawl like serpents upon their bellies, others have little fleshy tubercles which in some measure perform the office of feet: in the cuttle-fish, the long processes which surround the head perform the office of arms, feet, and fins: while the whole of the *Polypes*, or corals, with many other groups, are deprived of all power of moving from the spot in which they were born. These several peculiarities enable us to frame essential characters for accurate discrimination, of the most valuable description, because they are not only obvious, but keep together the individuals of small but natural assemblages. External are always better than internal distinctions; for it is surely more desirable that we should define an animal by something which every body can see, than to search for distinctions in its complex anatomical structure.

(188.) Before concluding this chapter, we shall offer a few remarks upon metamorphosis, in relation to the value of characters derived exclusively from its different variations. The early writers on natural history, previous to the time of Linnæus, attached so much consequence to the transformations which the insect world underwent, that their classifications were mainly founded thereon. Linnæus, for what precise reason does not exactly appear, decided on drawing the characters of his groups from the perfect insect alone; probably considering, and we think justly, that distinctions founded upon animals in their perfect state of existence, are more permanent and valuable than such as are taken from their immatured structure. Be this, however, as it may, metamorphosis, of late years, has been again brought into notice, by one of the first entomologists of the age, whose theory on the natural arrangement of the insect kingdom (*Annulosa*) is mainly founded on the mode of its variation. There can be no doubt of the truth of the two propositions laid down by Mr. M'Leay: first, that metamorphosis is the grand distinction of the *Annulosa;* and, secondly, that the mode of its variation will indicate the natural arrangement of the whole of the animals composing that class. Yet, while we admit the truth of this theory, we dissent from the application that is made of it. Metamorphosis, like all other characters, must not be made to violate nature by the separation of naturally connected groups. For, the moment we do this, we should suspect we are straining the lock, or have got the wrong key. If the first proposition just mentioned be correct, the logical inference will be, that those insects which exhibit the most distinct and striking transformations are consequently the *most* typical of all the annulose animals ; and that, in proportion as the metamorphosis of the rest is more or less perfect, so are the orders containing them removed from the typical pre-eminence. So far, however, from attempting, at the very onset, to demonstrate the truth of this proposition, by pointing out the most typical order of the *Annulosa*, Mr. M'Leay candidly confesses his inability so to do; thus failing to establish his theory, in that particular instance where its demonstration is most essential. This oversight, we trust to make it subsequently appear, has entirely arisen from his not following up the theoretical deduction he had come to on the value of metamorphosis: for, instead of founding his primary divisions upon it, he unfortunately adopted those of Clairville, taken from the mode of imbibing their food, and hence named *Mandibulata* and *Haustellata ;* thus, in fact, virtually denying the truth of the proposition assumed, and making the mode of taking food,— not metamorphosis,— the grand character upon which the primary divisions of the *Ptilota* repose. So acute an observer could not, however, fail to perceive the numberless difficulties which this error produced in his details; and prompted by his love of truth and nature, he makes no scruple to confess them on many occasions: nay, he candidly admits that he has not yet discovered the natural arrangement of the annulose circle, although every one must admit that his views on its relative affinities are any thing but artificial. Metamorphosis, in fact, is really one of the primary distinctions of the typical *Annulosa*, but it is not the only one; so, also, is the structure of the mouth. Yet neither of these, by themselves, will completely designate the typical groups. We know, by experience, that every peculiarity or variation in metamorphosis is almost always accompanied by external differences in structure, permanent in themselves, and always within reach of observation. Why, therefore, should we designate our groups by characters which are evanescent, when the same object can be attained by using others that are permanent? How is a student, for example, to discover the natural tribe to which any particular beetle belongs, and of whose metamorphosis he is entirely ignorant, if the tribes are to be characterised alone by their metamorphosis—that is, by the form of their caterpillar and chrysalis? The thing is manifestly impossible. But the evils of assuming this theory as infallible do not stop here. One of its most able and intelligent advocates has made metamorphosis the basis of his arrangement of the *Lepidoptera*; so that, if this plan be generally adopted, we shall never feel certain on the natural affinities of an insect, until we have studied its larva and pupa. For our own part, we must confess that we have the greatest objection to such characters; and we think that it is the duty of every naturalist to simplify the acquisition of science, by choosing such characters for his groups, as can be easily understood, and at all times verified. Now this cannot, of course, be said of such as are exclusively founded upon metamorphosis; and we therefore consider that metamorphosis, however valuable in helping to distinguish large assemblies, is pushed much further than nature warrants, when it is used as the chief corner stone for the construction of genera and sub-genera; which groups, if they are really natural, will always be distinguished by other and more intelligible characters.

(189.) From what has been just said, it must not, however, be inferred, that the metamorphosis of insects is to be disregarded; or that characters derived therefrom are not to be employed. The value of this and every other character depends upon the judicious skill with which it is used. In a group of unarranged animals, we can never know, *à priori*, what are truly its essential characters. It therefore becomes necessary to study all, that we may discover their relative prevalence, and thereupon make our selection. Another important advantage will also result from such investigations: some characters will be brought to light, which, although not employed as essential distinctions of the group, will nevertheless throw considerable light on its analogical relations. A more beautiful instance of this cannot possibly be found, than in the subordinate types of the genera *Amphrisius* and *Papilio*, detailed at some length in our *Zool. Il.* second series. Each of these genera forms a circular group; and the contents of one intimately correspond with the contents of the other, not only in the form of the perfect insect, but even in the larva of all such as are yet known. It has been well said, and no axiom requires to be more impressed upon the mind of the true follower of science, that a natural arrangement will stand any test. The two groups in question may be characterised by their external forms; and yet we see that the arrangement thus produced and founded on the perfect insect, developes the same theory of the variation of metamorphosis which pervades the whole of the order.

(190.) We have now attempted to point out the chief considerations which should influence the naturalist in his choice of characters; whereby he may define, with brevity and perspicuity, the numerous groups into which nature has divided the animal kingdom; or which it is necessary for us to keep in distinct allotments, until their true station in the scale of being is better understood. Characters, founded upon the circumstances here noted, are independent of all theory; inasmuch as they will repose upon facts of structure or of economy, which, in any system, must be kept distinct. In what sense they are to be used, or rather, to what description of groups they are applicable, is another question, which can only be solved by great experience, and by understanding the principles of variation in the different classes of the animal kingdom. Our present business is merely to point them out as solid materials for effecting scientific arrangement, leaving their application to the judgment and prudence of those who are competent to use them. Our own views, resulting from an attentive consideration of all these phenomena, will form the subject of a succeeding volume. We are not now either explaining or advocating any partiticular theory; but rather endeavouring to fix the attention upon circumstances which must enter into all theories. In every attempt to discover the natural system, be that system what it may, these matters cannot and must not be overlooked, if our desire is to discover and explain those general laws of nature by which all these diversities and resemblances can be reconciled and accounted for.

CHAP. VI.

ON THE IMPORTANCE OF ANALOGY WHEN APPLIED TO THE CONFIRMATION OF THEORY.

(191.) WE have already explained, and familiarly illustrated, the two sorts of relations which natural objects bear to each other, and which are distinguished by the terms *Analogy* and *Affinity*. The prevalence of these relations is so universal throughout nature, that there is no group of beings, however small, which does not present them. Nay, we question not that every individual species has its analogies, as it certainly must have its affinities. In a future volume we propose to enquire more particularly into these relations, and to bring forward such instances of their prevalence, as to sanction the hypothesis that they are uniform, constant, and universal in every part of the animal creation. In regard to affinities, indeed, this truth is self-evident; because, whatever forms part of a series, must of necessity have affinities, and these must be of different degrees. But, in regard to analogy, the case is different, and calls for a much more extended enquiry. On the present occasion, however, we shall merely consider those arguments which may be used, *à priori*, in favour of the supposition that analogies are, in the most comprehensive sense of the word, universal; and that they consequently assume an importance of the highest order when applied to illustrate, and to confirm, any theory on the variation of animal structure.

(192.) In the first place, it is unnecessary to enforce the axiom long established by sound philosophy, that natural and moral truths are but parts of the great system of nature. Nor need we go over those arguments that have been already so ably and so powerfully urged by others, to show that every thing in this world is evidently intended to be the means of moral and intellectual improvement, to a creature made capable of perceiving in it this use.* This perfect analogy between the moral and the natural world, no Christian in these days will even think of questioning, much less of disputing. It therefore follows, that as the material system of the universe possesses, as a whole, analogical properties, we are authorised in concluding that its several parts partake of the same nature as the whole, or, in other words, that the system upon which organised beings have been formed, — and which, by naturalists, is more especially termed the natural system, — possesses within itself that perfect exemplification of analogical relations, which we admit to exist between the natural and the moral worlds.

(193.) The greater the universality of any known law of nature is found to be, the more important does it become to the investigations of physical science. Between material and immaterial things, there is no other relation than that which is afforded by analogy: without this, they would be widely and totally distinct; with this, they are united; and one reciprocally illustrates the other. Analogy, or

---

* Hampden's Essay on the Philosophical Evidence of Christianity, xvii. See also the admirable volumes of Harness, " On the Connection of Christianity with Human Happiness."

symbolical representation, is, therefore, the most universal law of nature, because it embraces and extends its influence over the natural, the moral, and the spiritual world; a property which no other law, yet discovered, is known to possess. Hence we may infer that, in its more restricted application to natural history, it is equally paramount; and that, to this science, it is what the law of gravitation is known to be to astronomy.

(194.) It was, no doubt, from a perception of the vast importance of analogy, that the immortal Bacon so strongly recommends it in the investigation of nature; when, among other things which demand our attention, he enumerates, "*parallelas, sive similitudines physicas*," and, as an admirable reasoner on the same subject has happily stated*, having adverted to the practice of former philosophers in noting and explaining the actual differences among natural productions, as of little real use in constituting the sciences, he requires that pains should be bestowed rather in enquiring into, and noting, the similitudes and analogies of things; adding, at the same time, the just caution, that the similitudes should not be fortuitous and fanciful, but be real and substantial, and merged into nature.

(195.) The importance of analogical reasoning as a medium of proof, has been no less inculcated by one of the most profound philosophers of modern days. Dugald Stewart, in adverting to the opinions of Reid and Campbell on this subject, expresses his doubts whether both of these ingenious writers have not somewhat underrated the importance of analogy

---

* Hampden, p. 107.

as a medium of proof, and as a source of new information. " I acknowledge, at the same time," he continues, " that between the positive and the negative applications of this species of evidence, there is an essential difference. When employed to refute an objection, it may often furnish an argument irresistably and unanswerably convincing ; when employed as a medium of proof, it can never authorise more than a probable conjecture, inviting and encouraging further examination." But, as if sensible that, this latter assertion took from analogy its due weight, he proceeds to qualify it by adding, that " in some instances, however, the probability resulting from a concurrence of different analogies may rise so high as to produce an effect on the belief scarcely distinguishable from moral certainty."

(196.) Analogy, as it has been justly remarked, as a ground of illustration, is not essentially distinct from analogy as a ground of reasoning. For some may be disposed fully to concede the illustrative use of an appeal to the analogy of the moral and the natural world, as a means of conciliating a favourable hearing to the philosophy of zoology, but dispute the argumentative validity (and conclusiveness) of such an appeal. It should be observed, then, that unless that which purports to be an illustration of any thing, has a real foundation in nature for the comparison instituted, it cannot throw any true light on the subject to which it is applied. If the point of comparison be assumed, the application of the proposed illustration is only hypothetical; and the subject, in its proper nature, is rather obscured than enlightened by the false representation of it. Such, indeed, is the actual effect produced by fanciful analogies;—they darken the subject itself to which they are applied, whilst they diffuse over it their own specious colouring: hence such analogies, although fictitious, may be properly used by the poet or the orator, to ennoble and beautify subjects which require dignity and ornament; but they cannot for a moment be admitted into the precincts of physical science. An instance, indeed, in ordinary cases, on which a just analogy is founded, may in itself be fictitious, as in the employment of parables and fables, or in putting a supposed case; yet such instances, where science is out of the question, may be just analogies, because they are instances of some real principle obtained by previous induction, or actual observations embodied in some arbitrary form. They are, in fact, *latent inductions,* or philosophical truths divested of their proper evidence. The real difference, then, between an argumentative and an illustrative analogy, each being considered simply as such, consists in the form in which they are discerned. If each of several particulars analogically compared is otherwise known, and they are only brought together by analogy, then they are illustrations only of each other. But if certain particulars only are known, and these are employed for the investigation of another particular, then are the known particulars, *arguments* to the unknown one. The process, however, of detecting the justness of the analogy is the same in both cases.

(197.) Analogy is in all subjects the life and soul of illustration. It represents to us the same general truth under different forms, and under different points of view; and this property is in itself a fruitful source of instruction. For though the facts themselves which it connects, may be equally knowable in themselves, it does not follow that they are equally so to different minds. A simple truth, which to a particular mind seems isolated in itself, may be powerfully reflected upon by another truth, which peculiar habits of thought in that mind have rendered more familiar. Thus, if a divine was told that the progression of natural affinities, in any given group of animals, was in a circle, he might at first consider it, not being himself a naturalist, as a simple fact belonging only to zoological science; but if he reflected a moment on the subject, other truths with which he was more familiar would arise in his mind: it would occur to him, that the life of man, the course of the seasons, and the motion of the heavenly bodies, had their progression on the same principle; and that these were but types and shadows of that immense circle of eternity, which has had no beginning, and will have no end. With these truths he is familiar; with the former he was not: but applying the one to the other, he sees their mutual relations of analogy; and that which at first appeared to him an isolated fact, or an admitted truth, disconnected with those he was accustomed to contemplate, becomes irradiated with a flood of light, which is again reflected upon those truths which have been instrumental in enabling him to discern the vast extent of a simple law of nature.

(198.) From the tacit conviction of the uniformity of truth which every reflecting mind has acquired, we cannot be satisfied to see a truth unfolded to our apprehension in a single instance only, but we desire to perceive the instruction conveyed by any particular fact, depicted also in another instance, differing in some essential respects from that already before us; so that, from the various lights of different facts, concentrated in the point in question, we may form a correct judgment, whether the conclusion obtained from the first instance be a real principle of nature. If, for instance, any truth of anatomical science, collected from observation of our own species, were discerned also in the structure of most of the vertebrated animals, we should be almost sure that it was a general principle in that division of nature; but if, pursuing our examination into the invertebrated animals, we discovered the same principle under a different modification, and were enabled to trace all the intermediate steps of gradation between the two extremes, we should then be sure that the principle was a general law of nature; since we found that it held also where the peculiar circumstances, in which it was first observed, were wanting.

(199.) The variety which is introduced into any subjects by analogical argument, is in itself greatly serviceable to the business of instruction; it throws over the subject an inviting garb of attractiveness, thus alluring the attention of the general reader, and keeping alive the interest of the student. For example, in the analogy just quoted, what a pleasing and delightful illustration is given — by the circular progression of the seasons — of the circular progression of beings in nature! both exhibited in friendly contrast with some of the greatest truths of the material and the spiritual world. How different are the analogical instances! and yet how harmonious! The mind, thus led to the acknowledgment of the uniformity of truth, the universality, the grandeur, and the simplicity of nature's laws, obtains, in the act of learning, a delightful relaxation from the continued pressure of abstract scientific or doctrinal instruction: it recreates itself in the contemplation of the revolution of the seasons, or the diurnal course of the earth, and yields itself up a willing convert to the truth, over which such loveliness and harmony is diffused. Further, while analogy appeals so forcibly to the pleasure of association, making us acquainted with new things by the reflection cast upon them from other things with which we have long been familiar,—it also unites in its effect, as a means of instruction, a pleasure akin to that produced by imitation in the fine arts. These accomplish their purpose, by exciting that admiration which arises from perceiving some effect observed in nature attained under an artificial mode of execution.* An analogous fact may, in like manner, be considered as an imitation, under a different form, of another fact to which it is analogous. It is a resemblance, as close as the nature of the subjects, to which they respectively belong, will admit. We are pleased accordingly with the detection of such a resemblance, formed, as it were, in spite of the real discrepance of the subjects. The unexpected conformity of the different instances excites our admiration, and disposes us to a ready acquiescence in the belief that such analogies are not fanciful, but founded on a general law of nature.

(200.) The force of conviction which analogy,

---

* Adam Smith's Works, vol. v. p. 245.

when skilfully and consistently employed, brings with it, is probably less owing to any other advantage resulting from its use, than to this in particular, that it invests the learner with the character of self-instructor. It holds up to him some acknowledged fact, in which, as in a mirror, he may behold the truth in question; and leaves him to deduce it, almost by observation rather than by reasoning, from that which is brought before him. The mind which is thus illumined, instead of being alienated by the dogmatism of its teacher, or repelled by an assumption of superiority on his part, recognises in its own former conviction the truth which is introduced under a new garb, and accepts it as a just extension of a conclusion in which it has already acquiesced. It seems, indeed, to be exerting an act of recollection, instead of making fresh acquisitions of knowledge. That false pride, which recoils from the humiliating confession of error, and renders the intellect obdurate against the better reason, is thus beguiled into compliance with the arguments of an opponent; and the mind, thus relieved of the burthen of resistance to the truth, seems to say in secret to itself (as Aristotle observes of the effect of metaphor in some instances), ὡς ἀληθῶς, ἐγὼ δ' ἥμαρτον, recanting its error, while it confesses the truth.*

(201.) Such are the general effects and advantages produced by analogy in the elucidation of truth. Things which in their essential nature are totally opposite, are found, on closer investigation, to possess mutual relations, and to be governed by the

---

Hampden, Essay on the Phil. Evid. of Christ., p. 211

same law. Hence we discover three sorts of analogies pervading the system of nature, in the widest and most exalted application of the term: the first regards the spiritual truths of revelation; the second, those which belong only to the moral system; while the third are drawn from the phenomena of the material world. It would be foreign to our present purpose, did we here recapitulate the chief arguments by which the first and the second of these analogies have been so frequently and effectually illustrated; but in reference to what has been already said on the connection of the truths of natural history with those of religion, we cannot withhold from the reader the sound and philosophic reasonings of an author whom we have already so largely quoted, and whose arguments are peculiarly valuable to our present purpose, as being those not of a naturalist, but of a deep and original thinker, who, in making them, seems unconscious how applicable they are to legitimate science.

(202.) In maintaining the first proposition just stated, viz. that analogy establishes a connection between the truths of Scripture and the facts of nature, our author justly remarks that in so doing we must " refer each to that state of things with which it is immediately connected. We must examine whether, when all those circumstances which may naturally be supposed to produce the observed difference in the actual developement of theological truth, according as it belongs to the system of nature or that of grace, are taken into our consideration, the same abstract truth emerges as the point of ultimate coincidence. For if nothing appears to prevent such an ultimate coincidence of a fact of nature and a scriptural truth, but the peculiar circumstances of the two systems to which they respectively belong, it is evident that the two may justly be conceived as ultimately coinciding in principle, since they then appear as therefore only not coincident actually, because their circumstances are not. Hence it is, that the credibility derived to the Scriptures from the coincidence of their doctrines and circumstances with the facts of nature, is that which belongs to the evidence of analogy. For by analogical reasoning we are enabled to make the requisite references to the circumstances by which a general truth may be variously modified, and to express the result of such references in our conclusion. When we argue by induction, the conclusion embraces all the circumstances belonging to the facts upon which our observations have been made. We reject and exclude all that are merely accidental, but we rigidly preserve in the general proposition every particular which appears really to belong to the effect produced. Whenever, therefore, any circumstance really important is varied, our former induction fails, and we must then either repeat the experiment, or, if actual experiment be impracticable, we must have recourse to analogical reasoning; that is, to a mode of reasoning which affirms the conclusion with such reserves, such alterations, or exceptions, as may arise from any difference in the circumstances to which it is extended. Without, indeed, such a relative adaptation to the general truth as obtained by induction to the altered circumstances of the case, the inference would be evidently unsound, as appears from this consideration alone, — that, as every induction is relative to the circumstances under which it is made, and as analogy is only a substitute for induction, so also must analogy express, or at least imply, that relation to the altered circumstances which would have been expressed, had the conclusion been directly obtained by induction. And whether we are able to state exactly the effect which these new circumstances may produce, or can only allow for it by an implied reference to them, the conclusion is equally logical, since in either case we do not proceed beyond the limits of the premises. It remains, however, to ascertain whether the allowances which we make for the peculiarities of each mode of divine instruction, in tracing out by analysis the common principles into which they are ultimately resolvable, may be identified with those variations in the facts of each system, which might be anticipated in reasoning by analogy concerning the truths of Scripture, from the data furnished by experience. For this is necessary, in order to show that the facts of nature, and the doctrines of Scripture, are really analogous to each other. If the difference between a scriptural truth, and its counterpart in the system of nature, were greater or less than such as might be attributed to the difference of circumstances, the scriptural truth could not, in such a case, be regarded as a conclusion from experience, nor could the Christian religion be established as philosophically true.*

(203.) "While analogy is the happy instrument of

---

* Hampden, p. 56.

conveying light into subjects in general, it is peculiarly so when employed in elucidating the truths of religion. Here the force of contrast with which it acts is at the maximum. We bring together the things of heaven and the things of earth; and bestow on the most remote and inaccessible objects, some portion of that circumstantial particularity which belongs to those present and visible. To behold truths, in themselves so high above our comprehension, in connection with those which are familiarly inculcated on us by experience, must call forth our strongest admiration, and powerfully interests us on both sides, but particularly on that of our religion. Divine Wisdom then descends from its etherial seat, as the accessor of the throne of the Eternal, and communes with us face to face, and hand to hand. We find that the subjects of which Scripture treats are not chimeras, not creations of the fancy, which have no substantial existence; but things which *are:* things in which we live, and move, and have our being. It no longer appears to us in the light of a scheme, contrived in the bowers of philosophic seclusion, and addressing itself only to the contemplative and impassioned devotee, like the day dreams of the Koran, emerging from the gloom and solitude of the cave of Hara; but it shines forth conspicuously, as an energising principle, as a knowledge which is power, as a work of the Lord, carried on in the passing scene, with which we cannot help sympathising without doing violence to all the principles of our nature."*

---

* Hampden.

(204.) The power of analogy, thus so ably illustrated in the case of religion, is precisely as strong when applied to the elucidation and confirmation of a theory in physical science. When once what is supposed to be a general law of nature is discovered, its truth and certainty becomes more and more confirmed, in proportion to the variety and severity of the tests applied to it. And if, after tracing it, as widely as possible, through the almost numberless groups of the animal world, it becomes also apparent both in the vegetable and in the mineral kingdoms, we have all the evidence that human research or human wisdom can conceive, that our theory is sound; or, in other words, that we have achieved the discovery of one of those immutable truths which the wit of man can never devise, or the power of time destroy.

PART IV.

ON THE PRESENT STATE OF ZOOLOGICAL SCIENCE IN BRITAIN, AND ON THE MEANS BEST CALCULATED FOR ITS ENCOURAGEMENT AND EXTENSION.

CHAPTER I.

INTRODUCTORY REMARKS. — SOME ACCOUNT OF THE NATURE AND PRESENT STATE OF OUR SCIENTIFIC SOCIETIES AND INSTITUTIONS, AND OF THE MEANS THEY POSSESS OF ENCOURAGING SCIENCE. — NATIONAL ENCOURAGEMENT.

(205.) THE enquiry we are now to enter upon, although to some it may appear irrelevant, is yet intimately and vitally connected with the object of this volume. We have, in the preceding pages, laid before the reader those advantages — chiefly intellectual — which might allure him to the study of nature. He may, indeed, gather recreation and delight in limiting his contemplations to the simple objects which a rural walk affords to him. He may be content to admire a few detached ornaments of the temple, without desiring to understand the extent and harmonious construction of the building itself. But, if he desire to quit this humble path of enquiry for another more elevated, if he wish to generalise his ideas, and compare his observations with those of others, he is no longer, as in the former case, dependent upon his own resources; he must associate with those of similar pursuits and studies with himself. He must learn to distinguish that which is known from that which is unknown, and this can only be done by a reciprocal communication of knowledge. Hence, the origin of all societies. The value of such associations is greater, perhaps, than at first sight it appears to be; for, besides those advantages just mentioned, there is another, without which some of the most gifted minds would probably remain inert and inactive. Intercourse with congenial spirits excites that noble and generous emulation which has been the impelling principle of some of the greatest of men; and it will ever prompt them to the exertion of energies never before called into action. Many, therefore, of the best interests of science are involved in the construction of these societies; they exercise, in various ways, an important influence upon the advancement of human knowledge, and they consequently demand the serious consideration of those who feel interested in its extension. The present division of our treatise will, therefore, be devoted to this discussion. We shall, in the first place, take a hasty glance at the present state of the physical sciences generally, and of zoology in particular, with the view of ascertaining whether or not there exist adequate means of instruction or encouragement for its successful prosecution.

(206.) It cannot fail to be remarked, by those who watch the operations of the human mind, that the peculiarities of a nation may be traced in its public institutions. Nor, indeed, if we reflect but for a moment, can it be otherwise. Societies and associations, whatsoever may be their object, are the embodying into a tangible form, the private sentiments and feelings of individuals. Those who have lived sufficiently long among other nations to be disenchanted of their national prejudices, or who are disposed to believe the concurring testimony of the most enlightened foreigners, need hardly be told that, intermixed with many qualities of a nobler description, the English, essentially, are a proud, ostentatious, although a generous people. The two first of these characteristics are more prominent, perhaps, than the third. By which we mean to say, that there are more instances of ostentatious munificence, than of secret and disinterested generosity. If any one imagines this censure undeserved, we would only refer him to any list of charitable donations, where, for one contribution prefaced by the initials of a secret donor, he will find twenty blazoned forth with ostentatious parade. Now, the effects of these national characteristics are shown in our public institutions. From the union of generosity and ostentation, — sentiments fostered by our enormous wealth, — has sprung a greater number of charitable and benevolent associations than are to be met with in any three European nations. These noble institutions excite the admiration of every one who thinks upon the mass of human misery they tend to alleviate, and the incalculable good they disseminate to thousands, while they call forth the astonishment and praise of the surrounding nations. *Here*, our besetting sin of pride does not enter; the meanest as well as the highest are invited to join in the good work; and no man, because he has contributed largely to a charitable institution, or is one of those who presides as director or trustee over its affairs, ever dreams of signifying that honour to the world, by affixing some of the letters of the alphabet after his name.

(207.) But in our literary and scientific institutions the case is different. These, with few exceptions, have too much of an aristocratic character. We speak now of the leading societies of the metropolis; those from which, as chartered associations, we may gather the prevalent feeling. That love and respect for *wealth*, in the abstract, which forms so striking and so humiliating a blot on the national character, is no where more conspicuous than in one or two circumstances connected with these institutions. This is the more remarkable, because the blame attaches to those who, from a superior taste for intellectual pursuits, might be supposed exempt from the national idolatry of the vulgar. It is customary, indeed, to call the world of science a *republic*,—meaning thereby, as we presume, that all adventitious superiority resulting alone from wealth or rank, gives place to mental acquirements. But is such really the case? or, at least, is the principle itself really acted upon? No one can maintain that it is, when the fact is considered, that, with one solitary exception, all who wish to join these societies must contribute an annual payment. We can dispense with science in a candidate, but we must have his money. This is the plain, but the undeniable, fact. The exception above alluded to reflects the highest honour upon the oldest and the best of our societies, namely, the Linnæan Society of London. By the admission of associates, who are allowed to participate in all those discussions and proceedings of the members, which are purely of a scientific character, we give advantages to a highly deserving class of students, whose love for science may be equal to our own, but whose limited means preclude them from contributing pecuniary aid to the advancement of their favourite pursuits. The mutual advantages resulting from such a coalition need hardly be adverted to. On the one part, information on particular facts may be communicated, of the highest importance to the generalisations of the philosophic members; while, on the other hand, the mind of the practical investigator of nature will be improved and expanded by intercourse with those whom he will look up to as his masters, and whose society could never have been enjoyed, but for the removal of those barriers with which, in England, an undue regard to wealth has securely fenced the different grades of society.

(208.) The other peculiarity of our scientific institutions is, perhaps, more remarkable than the last, and equally serves to illustrate what has been expressed at the commencement of this chapter. Notwithstanding the number of our societies and associations, respectively formed for the advancement either of physical science in general, or any one of its numerous branches, there is not one of such a nature as to confer a *purely honourable distinction* on those whose pre-eminent abilities have placed them at the head of that particular science they cultivate. Taking, for instance, the Royal Society as the parent from which nearly all others have sprung, and which the public in general considers the most honourable, we find no classification whatever of the heterogeneous materials of which it is composed. Not the slightest distinction is made between the man of wealth, who pays his money to gratify his vanity; the mere dilettante, who feels a pleasure in the labours of others; or the accomplished philosopher, whose name may be celebrated throughout the world. All these, and many others, are admitted upon the same terms; all must pay an equal subscription, and all rank alike. Such, we believe, is universally the case with all the metropolitan societies. And in proportion as these have multiplied, so have the inconveniences arisen, not to say the insuperable obstacles, of concentrating the *élite* of science. If there existed a society to which no one could belong who did not possess scientific acquirements of known and acknowledged merit, it would matter but little if it entailed upon its members an annual contribution; because, although payment would be one of the requisites, it would not be the first, or the only one: the admission in itself would be an honour; because it would be placed beyond the reach of the most wealthy pretender, and would at once attach importance to a name. But, in the present state of our societies, all the qualifications expressed in the certificate of a candidate are, that "he is desirous of becoming a member, and likely to be a useful and valuable one:" the two latter requisites being generally interpreted, that he will promote the objects of the society, by paying his contributions regularly. So far as our own experience goes, the "mode of becoming a fellow" of the Royal Society, stated by Professor Babbage, is applicable to nearly all. " A. B. gets any three Fellows to sign a certificate, stating that he (A. B.) is desirous of becoming a member, and likely to be a useful and valuable one. This is handed in to the secretary, and suspended in the meeting-room. At the end of ten (or more) weeks, if A. B. has the good fortune to be perfectly unknown by any literary or scientific achievement, however small, he is quite sure of being elected, as a matter of course. If, on the other hand, he has unfortunately written on any subject connected with science, or is supposed to be acquainted with any branch of it, the members begin to enquire what he has done to deserve the honour? and, unless he has powerful friends, he has a fair chance of being black-balled. In fourteen years' experience," continues the same writer, "the few whom I have seen rejected, have all been known persons."*—(*Babbage, Reflections,* p. 51.)

(209.) This facility of acquiring diplomas is unquestionably one of the characteristics of our scientific institutions; and the evils which are the natural result, are already becoming apparent. An honorary distinction, when the qualifications upon which it was originally founded are lost sight of, so that it

---

* A singular verification of this occurred at one of the very few meetings at which, of late years, I have attended. It was the case of Captain P——, well known for his nautical discoveries and inventions. The same show of opposition was manifested at the election of a well known ornithological painter. But in both instances, by timely exertion, this strange opposition was defeated.

can be bought for money, becomes, of course, an intimation only of wealth, while its unlimited extension will soon render it insignificant. In proportion to the largeness of the subscription, will be the exclusion of those men whose names would add dignity to the list; for it has been truly observed, that " very few, indeed, of the cultivators of science rank amongst the wealthy classes;" while, in an inverse ratio, will be the admission of the titled and the untitled aristocracy. We cannot be persuaded that these predictions are exaggerated; because they are, from our personal knowledge, already in operation. Were it necessary to prove this, we could mention three or four names, whose fame has spread over the civilised world; but who, for the very reasons above mentioned, decline to become members of these aristocratic societies. Professor Babbage, in alluding to this subject, has given us a table of the admission fees payable to thirteen of the principal societies in Great Britain, with the appended letters, or " tail-pieces," attached to the names of the purchasers. " Thus," he continues, " those who are ambitious of scientific distinction, may, according to their fancy, render their name a kind of comet, carrying with it a tail of upwards of forty letters, at the average cost of $10l.$ $9s.$ $9\frac{1}{4}d.$ per letter."

(210.) Let us not, however, be misunderstood on this matter. It is far from our purpose, while we venture to point out the defects of these associations, to withhold the admission of their great usefulness. If perfection cannot be attained by an individual, how much less can it be expected in a corporate body? The only truths we wish to impress upon the reader, and of which he must be now convinced, are these: — 1. That the majority of our societies for the promotion of science are mainly, if not exclusively, formed for the wealthy; — that they are, essentially, aristocratic: 2. That, at this time, there is no mode of distinguishing, from their lists, such as are nominal from such as are really working members: and, 3. That the honour which once accompanied the admission into such societies has much deteriorated. Such are the nature and constitution of our London scientific bodies. We shall subsequently enquire how these defects can be remedied.

(211.) The most important object of these associations is the publication of essays or papers contributed by the members, and read at their meetings. These, in general, are collected into an annual volume, which is generally *distributed* to the members, and sold to the public. By this plan, the most effectual means are employed for disseminating knowledge at no extraordinary expense to the author. For it unfortunately is too true, that were he disposed to give his investigations to the world " at his own cost and charges," so little do the public estimate any work of pure science, that certain pecuniary loss to himself would be the result. It would be inconvenient and improper, on many accounts, for a society to publish *every* paper so communicated. It is, therefore, the business of the council to select the best of such as have been read; and if this is done faithfully and impartially, no better plan can possibly be pursued. But the imperfections of human nature will break out. It not unfrequently happens that partiality or prejudice enters largely into these decisions; for the publishing committee being irresponsible, their decision is final, and papers of the highest interest have been known to share the fate of " Rejected Addresses," when the views or theories they were intended to promulgate were in opposition to those entertained by the presiding judges.* By feelings of an opposite tendency, performances of little interest and less ability contrive to " pass muster," solely from the interest of friends " at the board." Hence it is generally found, that those members, who have the means of bringing their investigations before the scientific world through any other channel, are but scanty contributors to the transactions of societies; or, if they occasionally venture to send in a paper, take care that it should be confined to a matter-of-fact subject,—the plumage of a new bird, or the character of a new shell, — upon which there cannot, in ordinary cases, be a difference of opinion. Interest, in fact, is frequently as necessary for one description of these papers as for the other; for, otherwise, it is impossible to account for the insertion of one hundred and nine contributions in the *Philosophical Transactions* by a late eminent surgeon, many of which are not only erroneous in theory, but incorrect as to facts.

(212.) Oral discussion is limited to very few of our societies, although it is perhaps the most agree-

---

* A curious instance of this, furnished by the Zoological Society (the least scientific in its objects of all those in London), is mentioned in the Entomological Magazine, vol. ii. for January, 1834.

able and instructive mode of enlivening the proverbial heaviness of their meetings. It has been well said of such discussions, that besides the agreeable variation they create in the proceedings of the evening, they frequently bring together isolated facts in the science, which, however insignificant in themselves, mutually illustrate each other, and ultimately lead to important conclusions. The Geological Society, young, indeed, yet with all the vigour of manhood, is particularly celebrated in this respect; and others, we believe, of still more modern date, are following the example. It would be desirable if those of maturer years would conform to an innovation so well calculated to soften the dull austerity of their meetings.

(213.) The bestowal of medals, as a reward for high scientific investigations or discoveries, is at present confined, we believe, to two of these bodies, namely, the Society of Arts, and the Royal Society of Great Britain. The first of these comes not within our province; but, in discussing the subject before us, it becomes essential that we enter into some details regarding the latter. For these we must stand indebted to the information furnished by the *Reflections* of Professor Babbage; for, by a singularly injudicious remissness of very long standing, on the part of this society, no sort of information on these topics are given to the members, even upon their first admission; at least, in our own case, we were totally unacquainted with all these means of rewarding merit, and "*for exciting competition among men of science* *,"

---

* Vide Mr. Secretary Peel's Letter to the President of the Royal Society. — *Reflections*, p. 115.

possessed by the Royal Society, and which it might naturally have been supposed would be put into the hands of every member, to excite that " competition" which was the express condition upon which some of these medals were founded.

(214.) It appears, then, that the following means of rewarding merit are possessed by the president and council of the Royal Society. The first are the royal medals, two in number, of the value of fifty guineas each, founded by our late king, to be awarded annually as " honorary premiums, under the direction of the president and council, in such a manner as shall, by the excitement of competition among men of science, seem best calculated to promote the object for which the Royal Society was instituted."*
The following rules, for the award of these medals, were subsequently decided upon, by the council, on the 26th of January, 1826:—

> RESOLVED, That it is the opinion of the council that the medals be awarded for the most important discoveries or series of investigations, completed and made known to the Royal Society in the year preceding the day of their award. 2. That it is the opinion of the council that the presentation of the medals should not be limited to British subjects; and they propose, if it should be his Majesty's pleasure, that his effigy should form the obverse of the medal. 3. That two medals from the same die should be struck upon each foundation, one in gold, one in silver.

(215.) We neither possess the wish nor the means of enquiring how far these honorary rewards have been distributed with justice and impartiality. If what

---

* Mr. Secretary Peel's Letter.

has been asserted on this head by Professor Babbage is capable of satisfactory explanation, it was certainly incumbent on the parties concerned to have prevented the injurious influence of such statements, by giving such explanation. If, on the other hand, they cannot be contradicted, the society, as a body, must expect to suffer in public estimation. One thing, however, may be here observed, that we are unacquainted with any of our naturalists of the last century, on whom these distinctions have been conferred; although there are instances in former times of several having been so honoured.*

(216.) The remaining gifts for the promotion of science are also alone possessed by the Royal Society; they consist of pecuniary bequests in the shape of lectures; of which we believe there are but two, the Fairchild and the Croonian. The observations of Professor Babbage on the first of these are characterised by so much good sense and sound judgment, that we shall not weaken them by using different language, with the hope that the subject may claim that attention from the present council it so imperatively demands. It appears that a Mr. Fairchild, during the last century, " left by will twenty-five pounds to the Royal Society. This was increased by several subscriptions, and 100*l*. in the 3 per cent. South Sea annuities was purchased, the interest of which was to be devoted annually to pay for a sermon to be preached at St. Leonard's, Shoreditch. Few members of the society," observes our authority, " are aware, perhaps, either of the bequest or of its

---

* Ellis, Edwards, &c.

annual payment. I shall merely observe, that for five years, from 1800 to 1804, it was regularly given to Mr. Ascough, and that for twenty-six years it has been as regularly given to the Rev. Mr. E——. The annual amount is too trifling to stimulate to any extraordinary exertions; yet, small as it is, it might, if properly applied, be productive of much advantage to religion, and of great honour to the society. For this purpose it would be desirable that it should be delivered at some church or chapel more likely to be attended by members of the Royal Society. Notice of it should be given at the place of worship appointed, at least a week previous to its delivery, and at the two preceding weekly meetings of the Royal Society. The name of the gentleman nominated for that year, and the church at which the sermon should be preached, should be stated. With this publicity attending it, and by a judicious selection of the first two or three gentlemen appointed to deliver it, it would soon be esteemed an honour to be invited to compose such a lecture; and the society might always find, in its numerous list of members or aspirants, persons well qualified to fulfil a task as beneficial for the promotion of true religion as it ever must be for the interest of science. I am tempted to believe that such a course would call forth exertions of the most valuable character, as well as will give additional circulation to what is already done on that subject." Did these opinions stand in need of confirmation, we might appeal to the lectures delivered on this very foundation in 1784, and the three following years, by one of the soundest philosophers and most devout Christians that the history of our church can boast of; for the name of Jones of Nayland will ever shine as one of the brightest ornaments of the Christian profession. These discourses, to be found only among his works, illustrate, in the most simple and beautiful manner, many points of that harmonious analogy between the material and the spiritual world, between natural and revealed religion, which pervades creation. As a naturalist, this excellent man was not profound; for he lived when the philosophy of this science was in its infancy. How much more, then, could be achieved in these days by one who, like the present Woodwardian professor at Cambridge, could bring to the subject the richest stores of modern discovery, with the sound and orthodox principles of the established church.* On the Croonian lecture little need be said; it was instituted, according to Mr. Babbage, by Dr. Croone, for an annual essay on muscular motion. The payment, indeed, is but small, — three pounds, — yet still it might, like the last, be made a subject of honourable competition among medical students. At present, it seems to have been given, " as a sort of pension," year after year to one individual.

(217.) Such is the general nature of the chartered societies of this country formed for the promotion of science, and such are the means they possess

---

* Those who feel interested on this subject will peruse, with admiration and delight, the *Discourse on the Studies of the University of Cambridge*, by Professor Sedgwick; while the *Critique of Dr. Ure's Geology* in the British Review for July, 1829, by the same author, has been justly termed " an essay, equally worthy of a philosopher and a Christian."

of rewarding merit. There are some, however, of more recent origin, not yet formed into legal corporate bodies, and others of a mixed nature, which require separate consideration. It will be consequently necessary, in furtherance of our present object, to give the reader some particulars of the following, so far as they concern natural history, viz. the Royal, Linnæan, Geological, Zoological, Entomological.

(218.) The Royal Society, — if we are to judge from the contents of the printed *Transactions* of this body, the best criterion, perhaps, we can go by, — the Royal Society would seem to have almost banished every department of natural history (excepting that of comparative anatomy) from among the sciences which deserve their attention. At least, the papers occasionally to be found in these volumes, with few exceptions, are of a meagre and trivial nature. We know not whether this circumstance originates in the indifference of the council to such communications, or from the disinclination of those distinguished members, who cultivate this science, to hazard the rejection of their papers, or to see them lost, as it were, in a mass of others quite uncongenial. In former times — prior, indeed, to the institution of the Linnæan Society — natural history occupied a prominent place in these volumes; but such men as Ellis, Banks, and Solander have long passed away, and their successors in the same rank of science must be sought for in the continental academies. If this exclusion of zoological papers from the Royal Society's Transactions be really unintentional, it would be as well if some one of the present council by presenting such a paper would undeceive the members, more especially as such a circumstance might possibly induce others to follow the example.

(219.) The Linnæan Society of London, founded by the late amiable and excellent Sir James Smith, is not only the oldest, but the most honourable and efficient, of those devoted to natural history. From the date of its foundation, up to the present time, it has annually given to the world a volume of papers, unequal, indeed, in their respective merits, as such collections must inevitably be; but forming, as a whole, the most valuable collection of essays in this science in our language: it possesses an admirable library, partly formed by donations, and partly by purchase. Their museum, also, although chiefly rich in the birds of Australia (of which they possess the most valuable collection in this country), is by no means deficient in the productions of other regions, and has been more particularly enriched by the shells and insects of Sir Joseph Banks, the Indian birds of General Hardwicke, and the numerous presents of their late excellent and respected secretary, Alexander Macleay, Esq. now (to the great disadvantage of the society and of science at large) filling an arduous and honourable situation in the government of New Holland. All these scientific treasures are thrown open to the investigation not only of the members, but also of the *associates:* for it is to the honour of this society that such an accommodation is offered to those meritorious but unwealthy men to whom the payment of 35*l.* would be inconvenient; or who find it impossible to extend their pecuniary aid to all the societies in which they feel an interest. Here, also, the naturalist will be gratified by contemplating the entire library, museum, and herbarium of the celebrated man from whom the society takes its name. These collections, as it is well known, were purchased from the widow of Linné by Sir James Smith, during whose long and honourable career they were prodigiously augmented. The professional engagements of Sir James were not sufficiently lucrative to allow of his making a bequest of these treasures to that society he founded, and by which he was so much honoured and beloved. But the associated patriotism and disinterested liberality of the members accomplished that which it was not in the power of a single individual to do. Negotiations were opened with the trustees, and the purchase of the whole was at length effected. If there is any thing to be regretted in the construction of the Linnæan Society, it is the exclusion of oral discussions, the introduction of which, at no distant period, we hope will be effected.

(220.) The Geological Society is unquestionably the most active, and the most popular, even among scientific men, of all those which come under our notice. There is a vigour, an efficiency, and a liberality in all its proceedings which has called forth universal admiration; while the rapidity with which it has risen to its present eminence is the most convincing proof of the talents possessed by its leading members, and of the impartial manner in which its affairs are conducted. Of the objects pursued by this society, the only one which gives it a place in this list relates to the elucidation of fossil zoology, by investigating the nature of those animals whose remains lie buried in the accumulated strata of the globe, and which, in most cases, present us with forms of strange and unlooked-for structure. The oral discussions of the geologists are proverbial in the scientific world for the high intellectual gratification they usually give to visiters.

(221.) The constitution of the Zoological Society is of a very mixed nature, admirably adapted, indeed, to the reigning taste, and to uphold a very agreeable and popular establishment, suited to the rational amusement of the public. It is more calculated, however, to diffuse than to increase the actual stock of scientific knowledge. It possesses enormous funds; but it must not be forgotten, that for these funds it is largely indebted to its popular arrangements. It might perhaps combine, in a greater degree than it does, the diffusion of a taste for natural history with the permanent object of stimulating original investigation. The objects of this society are best expressed in the words of its prospectus; wherein "it is proposed to establish a society bearing the same relations to zoology and animal life that the Horticultural Society bears to botany and the vegetable kingdom. The object is to attempt the introduction of new races of quadrupeds, birds, fishes, &c. applicable to purposes of utility, either in our farm-yards, gardens, woods, waters, lakes, or rivers, and to connect with this object a general zoological collection of prepared specimens." In a subsequent notice it was intimated that a library would be attached to the museum. It is clear, however, from their present state, that the museum and the library are very secondary objects, although from the exhibition of the former, at one shilling each person, the public are tempted to believe it is much more extensive than others, which can be seen for nothing : it nevertheless derives much interest to the naturalist as containing the Sumatrian animals collected by Sir Stamford Raffles, and the North American birds described by Richardson and Swainson. Why these latter, collected at the public expense, and therefore public property, should not have been deposited in the National Museum, rather than have been given to a society of private individuals, is an enigma we cannot solve. That this donation was made without any regard to the interests of science is obvious from this simple fact, that, in their present situation, the specimens can only be seen by payment, nor can any scientific use be made of them but by permission of the council, and " at the discretion of the secretary:" whereas, had they gone to the British Museum, they might have been seen *gratis*, and used freely, without any such formal, tedious, and restricted regulations. The urbanity and liberal feeling of the secretary, indeed, is well calculated to diminish the inconvenience of debarring men of science from the free use of the materials it possesses. It is well known that admission to the privileges of a Fellow of the Zoological Society is a matter of no great difficulty. The forms of recommendation and election are observed, as in other societies ; but they are little more than forms. Upon the whole, however, the scientific character of the society, within the last two years, has much improved, and will doubtless continue so to do, as liberal feelings gain an ascendency in its councils. All who live near the Regent's Park, or who have the opportunity of enjoying the rational amusement to be found in the society's gardens, cannot do better than subscribe to its support. The collection of living animals is always interesting; and no expense, as we have been informed, is spared to provide a constant succession of novelties to attract the public. " The largeness of its income," observes Professor Babbage, " is a fearful consideration," but we have no desire to canvass its expenditure. We can only hope that a larger portion of these funds, in process of time, will be devoted to the prosecution and encouragement of legitimate science than has yet been done: the volume of *Transactions*, just published, may be considered a pledge that such will be the case. With such enormous funds, and with a judicious combination of science and of amusement, this society might eventually rank among the first in this or any other country.

(222.) The Surrey Zoological Gardens, although private property, are in no degree inferior to those in the Regent's Park, at least in regard to the number or variety of living animals. In this respect there is a sort of laudable rivalry between the two, very favourable to the gratification of the public. The Surrey possesses one advantage over its more aristocratic brother, highly important to the practical naturalist, who may go here to study, draw, or describe any animal in the collection, without encountering the advanced guard of illiberality in the shape of petitions, councils, secretaries, rules, and regulations. He has only to mention his wishes to

Mr. Cross, the highly respectable superintendent and chief proprietor, and he will not only receive immediate permission, but will have every information communicated to him which it is in the power of the attendants to furnish. Both gardens may be visited by the public on the payment of one shilling each person.

(223.) The Entomological Society is the youngest we have; for it was only founded in the autumn of the year 1833. The improved classifications of Dr. Leach, and the philosophic writings of the younger Macleay, first gave an impetus to this charming science, about fifteen years ago, which ever since has been slowly but progressively increasing. Nor must the fascinating volumes of Kirby and Spence be omitted, as contributing, even in a superior degree, to a more general diffusion of a love for entomology among a large portion of the intellectual classes. The fruits of this impetus are now beginning to ripen. A society of entomologists, young, ardent, and intelligent, has been formed, under the encouragement and support of their elder brethren; one of whom, alike distinguished for his love of science, his liberality of feeling, and his urbanity of manner, has been unanimously elected their efficient president. The constitution of the society is radically healthy: it has no titled officers elected only for their *name*. It is considered that science may be as well prosecuted without a charter of incorporation as with one: the contribution is small, and the members are effective. There is, in effect, no *quackery* in its composition. A society, so constituted, cannot fail to prosper, unless its council is so unwise as to plunge us into the expense of publishing *Transactions* at our " own charges," and thereby either involve the society in debt, or render it necessary to increase the subscription. As we have ventured to express our opinion upon this and other matters relative to the society elsewhere, it will be only necessary here to observe, that nearly all the entomologists of this country are among the members. Donations of books and specimens are sent from all quarters, so that a very good foundation already exists both for a library and museum. These are opened for inspection and use, unshackled by the official forms and delays so much complained of elsewhere. Oral discussions succeed the reading of more scientific information. Specimens of new or interesting subjects are exhibited and commented upon; and no mismanagement, as yet, has occasioned feuds or dissensions. Long may this state of things continue!

(224.) The chief of those societies established beyond the metropolis is the Philosophical Society of Cambridge, founded at that seat of learning about 1820, and incorporated by royal charter in the year 1832. It is not merely composed, as at first might be imagined, of the learned members of that university, but ranks among its fellows many of the most distinguished philosophers of Europe. The volumes of its *Transactions*, which have hitherto been given to the world, are second to none, published in this country, in the importance or the interest of their contents. Natural history essays, which at first were but thinly scattered, have increased in number in every succeeding volume; while the establishment of a museum and library, continually augmenting both by purchases and donations, have had a powerful effect in awakening the attention of the younger members to this most fascinating science: these sources of information, without which the student can make no effectual progress, are open to the inspection and use of every one. If we may entertain the hope that, at no very distant period, there will arise a school of *British* zoology, there are strong reasons for supposing its chief seat will be in our universities; and as that of Cambridge has long taken the lead in the cultivation of physical science, it seems highly probable that, with its present institutions, it will continue to maintain this distinction. So long as natural history was limited to a study of names, a comparison of specimens, and a discrimination of species, proficiency was always within the reach of the laborious compiler, the accumulating collector, and the minute discriminator. But so soon as labours of this sort ceased to assume more than a secondary importance, and they were discovered to be but instruments for attaining a higher degree of knowledge, from that moment the necessity will be felt of calling to our aid the exercise of superior mental faculties,—faculties which are rarely developed but by the expanding influence of an academic education. Hence it follows, that if we may expect to meet with such qualifications in any one particular class of the community, more than in another, they will be found among the students of our universities, early initiated in those sound philosophic principles which form the basis of human learning, and without which all sciences would become but a vast accumulation of isolated facts and disconnected conclusions: the grain of wheat and its useless chaff would be hoarded together, and the attention frittered and distracted by infinity of detail.

(225.) The sister university of Oxford has not yet, we believe, followed the noble example of Cambridge, by the institution of any particular society for cultivating zoology; yet it is not without a nucleus, around which it may be hoped, ere long, will be gathered a few ardent and enquiring spirits, which may eventually conquer the prejudices that still remain against the admission of natural history as a branch of academic education. Oxford will ever be associated in the mind of the naturalist with the names of those early promoters of our science, Tradescant and Ashmole, who, in an age of comparative ignorance, devoted their time and their fortune to the cultivation of natural history, and the formation of the earliest museum mentioned in our records. The remnant of this collection is now in existence, under the name of the *Ashmolean Museum*. As the reader will, doubtless, be gratified by the interesting account given of this museum by its present zealous curator, we shall here subjoin it for his perusal. " The museum presented to the university, and deposited in Oxford, in the year 1682, by Dr. Elias Ashmole, contained the first collection of objects illustrative of natural history which was ever formed in Britain; perhaps the first that was ever opened to public inspection in any nation of the whole world. This collection was made by the sagacity and industry of two ardent lovers of all that is beautiful and wonderful in nature, both named John Tradescant, a father and son. The father was, in 1629, gardener to King Charles I. The son travelled in North America, and imported new plants to the garden, and rarities to the museum, which was called the Ark, and duly visited by the dignified and the enlightened. The younger Tradescant bequeathed the museum, in 1662, to Ashmole, who was his friend, and the inmate of his house. The collection was certainly begun when natural science was in its infancy. Conrad Gesner, the illustrious and profoundly learned father of modern zoology, died in 1565. Aldrovandus had died poor and blind in the hospital of his native city, on which his learning conferred glory, in the year 1605. His works, however, together with those of Gesner, doubtless, gave stimulus and guidance to the labours of the Tradescants. Ray and Willughby were nearly contemporary with the son. If we suppose the elder Tradescant to have begun his collection in the year 1600, it will not be a subject of wonder that most of the skins of the animals should be in a state quite unserviceable to the purposes of science in the year 1824, when a renovation of this department of the museum was attempted. Several skins of fishes and reptiles, horns of African beasts, and bones of the elephant, the hippopotamus, and the grampus, still attest the well-directed ardour of the Tradescants. The legs and beaks of a few birds also are preserved, among which two deserve especial notice: one is the beak of the helmet hornbill, from the East Indies, which has been but lately imported in the entire state, having been long suspected to have been a foolish imposition contrived to deceive Tradescant; the other, the head of the dodo, or dodar, is the sole specimen existing of a bird larger than a swan, presented, probably by Mr. Thomas Herbert, to Tradescant, and brought by him from the island of Mauritius, where only it is reported to have been ever seen, and where it certainly does not now exist· That the stuffed skin was in the Tradescantian collection is proved by the catalogue, and by the incidental mention of it in Hyde's *Religio Veterum Persarum*, and by the statement of Ray."

(226.) The additions made to this private museum of the Tradescants by Ashmole were chiefly books. Borlace and Plot, — names connected with some of our most valuable county histories, — subsequently contributed to augment the foundation laid by Ashmole: and although a large portion of the animal preparations have long since crumbled into dust, the relics that remain are both interesting and valuable. The funds, however, left to maintain and enlarge this repository, are poor and inadequate. Ashmole left absolutely nothing to support the museum, while the profit of exhibition, ordained by his bequest, has rarely exceeded 30*l*. per annum. Dr. Rawlinson bequeathed 75*l*. to the keeper; but under the singular conditions that he should be an Englishman, not in orders, not a member of the Royal or Antiquarian Societies, &c. It is not improbable that the worthy Doctor, when he laid down these rules, shrewdly suspected that, without them, the place would become a sinecure given to some titled member of the university, who might have employed an illiterate deputy to perform that which required scientific attainments. Of late years, under the zealous and unremitting care of Mr. Duncan, the Ashmolean Museum, as we understand, has been re-arranged, and has received many valuable additions. We have been likewise assured, that a more general feeling in favour of the physical sciences pervades at Oxford than heretofore, and that there is a strong desire, among several influential members, to follow up the examples already exhibited at Cambridge. It must not be concealed, however, that this exclusion of zoology, as a " part and parcel" of our academic studies, is a national stigma: that it has repeatedly been adverted to, in terms of regret and of censure, by our own writers; and that it calls forth the astonishment and reproach of every enlightened foreigner. A stranger, ignorant of our national peculiarities, would almost imagine, from the rigour with which their study is enforced, that the writings of the heathen poets were peculiarly adapted to purify the heart, and curb the licentiousness of the youthful imagination; or that they formed, in some inexplicable way, a string of commentaries upon our religious creed.* And he might be further led to suppose that those wonders of the visible creation, which, when considered, will bring home conviction to the philosophic sceptic, were unworthy of study or regard, as if they were things of mere chance, — produced by a congregation of fortuitous atoms, alike incapable of demonstrating

---

* See the admirable remarks bearing on this subject in Forster's Essays, 8th ed. p. 348—374.

the being of a God, or the care He bestows upon his creatures.

(227.) Of our metropolitan institutions for public instruction we may be expected to say a few words. Two of these contain zoological chairs, but so poorly supported, that they excite little or no competition among those best able to give dignity and usefulness to their duties. Both these sources of instruction and enquiry are open to the naturalists of London, and may in time become highly important to the advancement of zoology.

(228.) The universities both of Edinburgh and of Glasgow have their Regius Professors of natural history, by whom lectures are given, and where museums of considerable extent are established. We possess, indeed, but little information on the actual state of zoological science in Scotland, further than what may be gathered from the " Transactions of the Wernerian Society ; " and the occasional papers inserted in that highly valuable publication, " The Edinburgh Philosophical Journal," of Professor Jameson. The museum of Glasgow has been described as already rich, and continually augmenting; while that of the Edinburgh College, so justly celebrated for the perfection of the specimens, seems to be second only to the national collection in the British Museum. The talents and capability of Professor Jameson, who fills the zoological chair in our northern capital, are well known, and have, doubtless, given an impetus to the science, which is even now beginning to show itself in cheap and popular compilations; precursors, let us hope, of something better. We regret the inability of giving, from personal knowledge, a more detailed account of these two establishments, which reflect so much honour upon our northern universities, and should long ago have been extended to those of England.

(229.) There are several provincial philosophical institutions and societies, either comprehending natural history as one of the sciences to be cultivated, or expressly devoted to it. The most important of these is the Natural History Society of Manchester, a town long and justly famed, not only for its commercial importance, but for its attachment to the physical sciences; an union so rare, that we know not where to find its parallel. The society in question has its periodical meetings, and is supported by the annual contributions of a very considerable number of members residing in that part of Lancashire. We have had the gratification of seeing what has been the result of this liberality; and we hesitate not to say that the zoological collection of this society, with a solitary exception, is second to none in the metropolis of Great Britain. Besides a very fine collection of native birds, it is rich in the ornithology of Tropical America and of the United States. The collection of insects is also extensive; but that of the *Testacea* yields only to the British Museum in the number, the rarity, and the interest of the specimens. The shells, in fact, amount to between 5000 and 6000 species; very many of which are undescribed, while others formed the chief ornaments of the Bligh, the Angus, and the Swainsonian collections. This fact proves that commercial and manufacturing occupations are by no means unfavourable to the prosecution of intellectual studies. This is apparent, not only in the higher and more educated classes of Manchester, but is very general through the operative class of the community. We were particularly struck one day during our visit, at seeing two or three individuals of the latter description attentively looking at some of the specimens in the museum, and comparing them with others brought for the purpose. The superior tone and manners of these humble admirers of nature are very striking, and at once show the effect of such tastes upon the inward man.

(230.) The opulent town of Liverpool, supposed by some to be superior in a commercial view, can bear no comparison with its neighbour in those intellectual pursuits of which we are speaking. There is, indeed, a Royal Institution, the schools of which, we have heard, are well conducted. We were encouraged, some years ago, to devote much time and trouble in the formation of a museum attached to this building; but, with the death of that illustrious historian to whose exertions and influence this town is chiefly indebted for its public institutions, expired that zeal for following up what had been so well begun. The museum has remained nearly in the same state, and, although admired, has not been adequately supported; while the taste for natural history, once very prevalent, has almost expired with the death of some, and the departure of others, whose intellectual superiority shed a lustre on the town of Liverpool. We admit, indeed, that in such a place, where almost every one is either engaged in commercial or professional pursuits, it cannot be expected that gentlemen will be found, who can afford to dedicate their time to the details of abstract science. The institutions of Bristol, and, as we believe, of Manchester, have seen this, and have accordingly appointed curators to their museums and libraries; the duties of which, quite distinct from those of the secretary, can only be advantageously performed by a person well versed in taxidermy, or the preservation of animals; and who has a competent knowledge of practical natural history in general. Such individuals are always to be found, but their remuneration must be proportionably adequate, and sufficient to render the situation respectable. A zoological garden has recently been established at Liverpool; which, as an ornament to the town, and a recreation to its inhabitants, will, doubtless, be much encouraged; and may in time lead to some useful and scientific purpose.

(231.) There is yet another institution, or rather society, of a higher and more comprehensive description than those just noticed, and which differs most materially from all the local associations we have yet noticed. We allude to the *British Association for the Advancement of Science*, instituted very recently, in imitation of that scientific congress of learned men upon the Continent, whose proceedings have become so celebrated. Some misapprehensions respecting the objects embraced by our British Association arose in the first instance, which deterred several of our *working savans* from immediately joining it: doubts also were entertained as to the expediency of contributing papers, and thereby impoverishing the *Transactions* of older societies, at a time when the higher walks of science were becoming nearly deserted. Happily, however, these doubts and fears have been dissipated, and the Association has rapidly grown into a large, influential, and energetic body. It is composed not only of men whose names are already known, but a very large number of juniors, particularly from the universities; and these have been joined by many influential individuals, attached to intellectual pursuits, who, by their countenance and support, uphold the cause of science in the eyes of the public. The terms of admission, therefore, are easy; and the pecuniary contribution very small. The proceedings of the Association differ materially from all those we have yet noticed. The meetings are annual, but, instead of being held at one place, the members assemble every year at some one of the great towns or cities of the empire. York, Oxford, Cambridge, and Edinburgh have already witnessed this intellectual jubilee; for such it may be truly called, since it brings together men of known reputation and of congenial pursuits, separated by distance from personal intercourse. It may readily be supposed that such meetings unite all the advantages of those held by the *stationary* societies, with many others *they* cannot possess; and that a spirit of excitement and of tempered conviviality enlivens the whole; giving to this assembling together of the votaries of science the charm and the relaxation of a holiday week. Yet there is still work to be done: the members are arranged into parties or *sections*, according to their respective pursuits; and Reports are drawn up by each, of the progress made in any particular branch of knowledge during the past year. The greatest hospitality is generally shown by such members as are resident, to those who come from a distance; speeches are made, toasts are drunk; and we can only regret the fate of those, who, from professional or other pursuits, have not the power of making so long an absence from home, and of sharing the intellectual and social pleasures of such instructive and pleasurable meetings.*

(232.) Having now dwelt at some length upon those aids and encouragements to science which emanate from public societies and institutes formed for that express purpose, we must be allowed to advert to another association, whose objects, indeed, are commercial, but whose patronage of science in all that relates to the civil and natural history of Asia is without parallel, and entitles THE HONOURABLE COMPANY OF MERCHANTS TRADING TO THE EAST INDIES not only to a place among the scientific institutions of this empire, but to rank with the first and foremost of those in Europe. We here look to this Company only in its connection

---

* This association originated in a suggestion of Sir David Brewster, who also took an active part in its subsequent organisation. Science is also indebted largely for its success to the unwearied zeal and incessant exertions of the Hon. and Rev. Vernon Harcourt, its general secretary, without whose aid it could scarcely have emerged from its infancy. It would be invidious to select for especial notice the names of other members, where so many are conspicuous; nevertheless we cannot omit that of Mr. Phillips, of York, the assistant secretary. — ED.

with the literature and science of the East. The liberality which the different courts of directors have shown, for a long series of years, in bringing to light the ancient records of that vast empire over which their authority extends, is attested by the publications these materials have given rise to, and the efficient patronage that has uniformly been extended to their authors. Every thing, in short, which could illustrate the ancient state of those singular nations now under the dominion of Britain, has been studiously sought for by the servants of the Company, and deposited in their archives. The Asiatic Society, celebrated for its learned Transactions since the days of Sir William Jones, owed its origin to their fostering care; while the splendid library and collection of Oriental MSS. at the India House attest the feelings which have so long prevaded their councils. If we turn, on the other hand, to what has been done for elucidating the natural history of their possessions, the result is still more conspicuous. A botanical garden, worthy of an Eastern monarch, superintended by distinguished botanists, having at their command all necessary assistants, has disseminated the splendours of the Indian flora over all similar establishments in Europe. Yet this liberality is not confined to public gardens, or to favoured botanists. Any individual of respectability, upon his return to Europe, may receive a collection of seeds and roots from these gardens, free of expense. Nor are these all the benefits resulting to the botanical world from the munificence of the Company. The different provinces of India have been explored by competent botanists; and thousands and tens of thousands of dried specimens, prepared under their superintendence, have been transmitted to England, arranged into separate collections, and then distributed among the scientific botanists of Europe. The same patronage has been extended to every thing regarding zoology. No sooner had the British arms taken possession of Java, than arrangements were made for securing the services of Dr. Horsfield, an eminent naturalist then residing in the island, and his valuable collections were made over to the Company. On the arrival of Dr. Horsfield in this country, these scientific treasures were deposited in the India House; and when suitable arrangements had been made in the museum for their reception, they were opened to the public and to men of science: and the " Zoological Researches in Java" were soon after published, under the Company's patronage. The chief results of Dr. Horsfield's discoveries being thus given to the world, the rich collection of duplicate specimens was ordered to be distributed, like those of the plants, among the different public museums, and the eminent zoologists, both in Britain and on the Continent. The splendid collection of insects, equally rich in duplicates, will, no doubt, be employed in a manner equally calculated to benefit science, so soon as the honour attached to their discovery and investigation has been secured. In short, in whatever light we view the scientific patronage exercised by the India Company, it is scarcely possible to do justice to that munificent spirit which is apparent in all the details.

(233.) From such an example of scientific patronage, emanating from a company of merchants, we turn with sorrow and regret to the next subject of discussion, namely, the support which zoology receives from the government of this country,—the most powerful and wealthy nation in Europe, whose pecuniary resources surpass all others, and whose profuse liberality, on almost all other subjects but those connected with science and art, nearly amounts to prodigality. We might, indeed, spare ourselves and our readers the humiliating detail which our present object imperatively demands, by summing up the whole with the confession that, excepting the British Museum, there is no national institution of any sort or kind for the teaching of natural history, for its prosecution or encouragement, or for the reward of its professors. Whether this indifference, or rather apathy, to a science so intimately connected as this is with religion, is expedient or politic, may be worth enquiry; but that it is unexampled among civilised nations, is a fact too notorious to be questioned. Before, however, we look further into this evil, let us take a general survey of the institution which forms the solitary exception above alluded to.

(234.) The British Museum is the repository for the national collections of books, manuscripts, sculptures, and natural productions. It owes its foundation to the purchase, made by parliament, of the entire collections of Sir Hans Sloane, President of the Royal Society for many years, and one of the most eminent physicians and patrons of learning of the age in which he lived. There is a curious document, published in a work * where no one would seek for it, regarding this famous collection, which we shall here insert.

An Account of the Names and Numbers of the several Species of Things contained in the Museum of Sir Hans Sloane, Bart., and which, since his Death, are placed for the Use of the Public in the British Museum.

The Library, including books of drawings, manuscripts, and prints, amounting to about - - - volumes	50,000
Medals and coins, ancient and modern - -	23,000
Cameos and intaglios, about - - -	700
Seals, &c. - - - -	268
Vessels, &c. of agate, jasper, &c. - -	542
Antiquities - - - -	1,125
Precious stones, agates, jaspers, &c. - -	2,256
Metals, minerals, ores, &c. - - -	2,725
Crystals, spars, &c. - - -	1,864
Fossils, flints, stones, &c. - -	1,275
Earths, sands, salts, &c. - - -	1,035
Bitumens, sulphurs, ambers, &c. - -	399
Talcs, mica, &c. - - - -	388
* Corals, sponges, &c. - - - -	1,421
* Testacea or shells - - - -	5,843
* Echini, Echinites, &c. - - -	659
* Asteriæ, Trochi, Entrochi, &c. - -	241
* Crustaceæ, crabs, lobsters, &c. - -	363

* Edwards's Gleanings of Natural History, vol. i. p. 7. Pref.

* Stellæ marinæ, star-fishes, &c. - - 173
* Fishes and their parts - - - 1,555
* Birds and their parts, eggs and nests of different species - - - - 1,172
* Quadrupeds - - - - - 1,886
* Vipers, serpents, &c. - - - 521
* Insects, &c. - - - - - 5,439
Vegetables - - - - 12,506
Hortus Siccus, or volumes of dried plants - 334
Humana, as calculi, anatomical preparations, &c. - - - - - 756
Miscellaneous things, natural, &c. - - 2,098
Mathematical instruments - - - 55
Pictures and drawings, framed - - 471

(235.) Edwards further adds, " Every single particular of all the above articles are numbered and entered by name, with short accounts of them, in thirty-eight volumes in folio, and eight in quarto." It would be an interesting enquiry to ascertain how many of the zoological subjects, originally in this vast Museum, are now in existence.† The total

---

† A few years ago, when the zoological collections of the Museum formed the subject of a debate in the House of Commons, and some censures were cast upon the little care *then* bestowed upon them, it was positively asserted by a ministerial member, since elevated to the peerage, that Sir H. Sloane's insects were *all in good preservation*. And this assertion was suffered to remain uncontradicted, from sheer ignorance in the opposition members, who appeared to know as little about the matter under discussion, as if it related to the Museum of China. The fact being, that no insects, as then preserved, could by any possibility, have existed so long.

number comprised under the separate items marked there being no less than 19,272. Nineteen thousand of these, in all probability, have perished, either from the imperfect manner in which they were prepared, or from the neglect which accompanied a long period of indifference, in the former conservators, to their proper custody. It is very probable that the leg of that extinct bird, the dodo, already adverted to, and now forming the greatest curiosity in the ornithological department, is one of the Sloanean relics. This noble basis having been laid, successive donations and purchases were added, and successive losses were suffered, as zeal or supineness on the part of succeeding curators predominated. This state of things continued until the appointment of Dr. Leach, a naturalist whose ability was only equalled by his zeal, and whose health eventually fell a sacrifice to incessant labour. Well do we remember the time when he set to work manfully in cleaning out what was then an Augean stable—a chaos of " confusion worse confounded." But the effects of long years of misrule and of disorder were not to be overcome by a single individual, who, while he was stopping the plague in one quarter, was necessitated to permit its full rage in another. Duties which, to be performed, would have required the activity of five or six naturalists, were imposed upon one; the task was Herculean, and, as his friends foresaw, he sunk under their burthen.

(236.) Whether this lamentable circumstance forced conviction upon the trustees, that the zoological department required augmentation, or whether the opening of the Continent, by showing us the state of other national museums, made the poverty of our own but too apparent, yet it was about this time that two assistants were appointed to aid the gentleman who next succeeded to the situation, and whose zeal in his official duties, and whose courteous demeanour to all who frequent the Museum, either for information or pleasure, have been so frequently praised: nor would it be just towards those who share these duties, to omit a public attestation of their promptitude in giving every facility in their power to all who require it, without that punctilious regard to those strict regulations which are certainly necessary, but which are sometimes highly inconvenient to students. Of the sums of money, worthily voted by the nation to this establishment, a large portion has been expended in erecting more suitable apartments for the natural history collections, the whole of which have been removed from the dark and dismal rooms they once occupied, and are now arranged, or arranging, in the new buildings. Were we to judge, indeed, merely from outward appearances, and compare the present state of the zoological collections of this Museum, with what they were ten years ago, we should be tempted to think that natural history was really patronised by the executive government; nor would it be supposed that so imposing an appearance could exist, with a deficiency of all those measures calculated to give proper efficacy to such an establishment. The only disadvantages of the zoological collections arise from the age and imperfect preservation of the ornithological specimens, and the poverty of the *Mammalia:* the shells are particularly fine, and the entomological cabinets, although poor in some of the orders, are very rich in others. The funds set apart for the purchase of additions, are, as it is understood, very scanty; but the public have began to be liberal in donations, for they see they are taken care of, and they naturally prefer sending them to the British Museum, where their gifts can be viewed gratuitously, than giving them to other collections, the managers of which oblige them to pay for seeing their own presents.

(237.) We must not omit, in this place, to notice two facts of recent occurrence, not so much from any great influence they have in themselves in retrieving the national character from the stigma of indifference to science, but that every indication, however slight, of an awakened sense to the importance of the subject, must be hailed with pleasure. We allude to some of the highest dignitaries of science having had bestowed upon them " the lowest title that is given to the lowest benefactor of the nation, or to the humblest servant of the crown *," and to the circumstance of one thousand pounds having been allotted by government to the execution of the zoological plates accompanying the volumes of the *Fauna Americani Boreali;* without this grant, indeed, the result of the zoological discoveries made by the Arctic expeditions of Franklin and Richardson would never have been given to the world. It will ever be an honour attached to the name of Lord Goderich, that he was the first minister of the

---

* Quarterly Review, No. 86.

crown who induced the British government to aid in the expense of publishing the fruits of zoological researches, carried on at the cost of the nation, the work in question being the first and nearly only instance of such liberality.

CHAP. II.

ON THE NATIONAL PATRONAGE OF SCIENCE IN OTHER COUNTRIES, AS COMPARED TO ITS NEGLECT BY THE BRITISH GOVERNMENT.—THE CAUSES WHICH PRODUCE THIS NEGLECT, AND THE EXPEDIENCY OF REMOVING THEM.

(238.) THE facts detailed in the last chapter being admitted, because they are notoriously undeniable, we come then to the following questions:— 1. Is such a state of things peculiar to England, or common to other nations? 2. Does abstract science, more particularly zoology, stand in need of any peculiar or national encouragement? and, 3. What are the causes which operate to its neglect? By discussing these questions impartially and dispassionately, we shall then be prepared to form a sound opinion, whether science, among us, is in a healthy state; and whether, in truth, it is advancing, quiescent, or retrograding. On a subject in every way so important and interesting, we have much fear, after the able manner in which it has already been treated, of not doing justice to the cause we advocate. But the general sense of any body of men can only be gathered from the expression of individual opinions; and although some of these will be more eloquent, and the reasons assigned more convincing, than others; yet, if they advocate the same general principles, and concur in the same sentiments, the reiterations of the weaker will give confirmation to the arguments of the more powerful advocate; and if several of these, taking up different departments of science, — each in their own walk, — arrive by different inferences at the same conclusions, we may safely believe that there is much of truth in the result. It may be said, indeed, by those who yet concur in the sentiments here expressed, that there is little need of any further discussion on the state of science in Britain, seeing it has already been animadverted upon, " more in sorrow than in anger," by such men as Sir H. Davy, Sir J. Herschel, Professor Babbage*, and Sir James South †; and further, by a writer no less accomplished than eloquent, in a Journal ‡ devoted to the political interests of the court or conservative party, and which would not have been the organ for casting imputations upon the government, except under strong and peculiar circumstances. With such a mass of evidence before those who have the power of remedying the evils complained of, it may be said, that to reiterate these complaints is alike tedious and unprofitable, seeing that they are already well known. But the question more properly is this: — Have they been redressed? have they made such an impression as they ought to make? have any effectual measures been taken

---

* Reflections on the Decline of Science in England. London, 1830.

† Charges against the President and Councils of the Royal Society. London, 1830.

‡ On the Decline of Science in England. Quarterly Review, Oct. 1830.

in consequence? or has any efficient reformation actually commenced? Until these questions can be answered more satisfactorily than at present, the more frequently such demands are urged, the more likely are they finally to receive attention, even from the very weariness of the complainants. It is not to be supposed that the aristocracy of science, proud even in their degradation, can derive either individual pleasure, or popular respect, in proclaiming to the world the little estimation in which they are held by the rulers of their country, — those, in fact, whose honourable duty it is to foster their exertions and reward their merit. They only seek to hold among the different grades of the national assemblies their proper rank and station, and equally to participate with others in those rewards and honours which should be the outward signs to the world at large of their intellectual merit. It is only, then, as a last resource, that they bring themselves to the humiliating alternative of public complaint, consoled by the reflection that, however those complaints for a time may be disregarded, yet that, if they are repeatedly made, a season will come when honest conviction will see their justice, and grant their demands. Furthermore, such statements should be more especially made in publications having a great circulation, as more likely to fix the attention of the public, and to come within the circle of those very few, in an exalted sphere of life, who have so much in their power to remedy what is amiss.

(239.) To arrive at a just conclusion on the questions before us, there is only one assertion which must be asssumed as true, — namely, that the greatness of a nation depends, not upon its physical, but on its intellectual power. From this axiom it necessarily follows that science produces manufactories, — manufactories commerce, and commerce wealth. Hence it is at once perceived how inseparably science is interwoven in all that gives power and dignity to a nation, or rather that it is the cornerstone upon which all other forms of greatness are built. The expediency, therefore, not to say the necessity — that physical knowledge should be nurtured and protected by every government, requires no discussion. It may, indeed, be urged by those who are fully aware of the connections just alluded to, that in a commercial country like this, all sciences which can be brought to bear upon the necessities, the conveniences, or the luxuries of life, require no other aid or reward than that which they are sure to meet with on their successful application. The moment that an invention is found really available for practical purposes, from that moment its author may fairly calculate upon receiving his reward by the general adoption of his discovery; while, on the other hand, if there be no real utility in the thing itself, the most powerful patronage will fail to establish it in public estimation. There is, undoubtedly, much of truth in these remarks, but a moment's reflection will convince us that they are partial and superficial. We merely notice them in this place, to apprise the reader that they have not been overlooked, and that they will form a point of separate discussion hereafter.

(240.) The first question, — Is science less

cultivated and held in less estimation in England than on the Continent?—is perhaps too general. A distinction must then be drawn between abstract and practical science; or, in other words, between such branches as can be brought to bear upon the physical wants and necessities of society, and such as are purely intellectual. There is still a further distinction, — and it is a very important one, — between that degree of knowledge which can render a science pleasing and popular, and that which, from aiming at the highest objects, the discovery of new laws, and the investigation of difficult questions, renders it, in the eyes of the many, abstruse and uninteresting. To each of these minor questions very different answers would be given, and they should therefore be considered separately. If we speak of science generally, it may fairly be questioned, whether, at any former period of our history it was ever held in so much estimation, or was so generally diffused among the mass of our countrymen, as it is at present. Yet, while we may truly exult in this awakening of the national intellect, we must remember that *diffusion* and *advancement* are two very different processes: and each may exist independent of the other. It is very essential, therefore, to our present purpose, when we speak of the *diffusion* or *extension* of science, that we do not confound these stages of developement with *discovery* or *advancement;* since the latter may be as different from the former as depth is from shallowness. Reverting, then, to the simple question, whether the higher walks of science, properly so called, are more neglected in England than on the Continent, there are few of our readers, we imagine, who will deny that any authorities can be greater than those just named, each of whom, speaking with reference to those sciences which they particularly study, concur in opinion that England has rapidly fallen behind the neighbouring nations. In reference to zoology, we have already given reasons in another place* for forming the same conclusion, nor need we stop there. We have long consigned comparative anatomy as regards the *Vertebrata*, as if by indolent consent, to the French; while that of the *Annulosa* has only just begun to excite some attention among us, in consequence of the splendid essays of Chablier, Leon Dufour, and others. Who, let us ask, has done any thing to elucidate the structure of the naked *Mollusca*, by following up the splendid discoveries of Savigny, the delicate and inimitable dissections of Poli, and a host of others, emanating from the zoological schools of France, Germany, and Italy? But this is not all: so little are the higher objects — the true philosophy — of our science, esteemed or cultivated, that discoveries of the first order, which open a new and unexpected field for the most important generalisations, and which will eventually overturn all the existing dogmas of systematists, these discoveries we have suffered to die almost in their birth, although they have actually been made by our own countrymen! Who, let us ask, has attempted to verify and follow up M'Leay's theory on the arrangement of the

---

* Northern Zoology, vol. ii. pref. p. xl.

Petalocera Lamillicornes, — insects which every one is ambitious of *collecting*, but which no one among us thinks of investigating? We have, again, the splendid discovery of Thompson on the metamorphosis of the *Crustacea* and of the *Cirripeda*. These discoveries, of which the last is worthy of Trembley or of Savigny, far from having been rewarded, as they deserved, by a Copleyan medal, have neither been investigated or verified; they were scarcely noticed in our journals, and although made within the last few years, they seem to be altogether forgotten! It is alike irksome and unnecessary to make further appeal to facts such as these, which verify too truly and too forcibly, the utter neglect of the philosophy of natural history in Britain, and this at the very time when frivolities of nomenclature and minutiæ of species occupy the attention of its followers, and when its commonplace facts and amusing details are dressed in popular language, published in every possible form, and perused with avidity. True it is that these cheap compilations and amusing collections of anecdotes have awakened a very general taste for natural history; and so far they are useful; but we are looking at present, not to the *extension*, whatever may be the ultimate result, but to the *advancement* of this science. And we unhesitatingly repeat, that its progress is more retrograde than otherwise.* If there is no taste for cultivating the higher investigations of zoology among those who are considered

---

* See North. Zool. vol. ii. pref. p. xliii.

its teachers, how can it be expected that what is not found in individuals should be found in learned societies? or how can it be imagined, under these circumstances, that the rising school of students should appreciate the value of those researches which alone give dignity to the study of nature?

(241.) From individuals and societies let us turn to national encouragement; that we may form some idea whether the governments of other nations regard science and its professors in the same light as they are viewed by that of Britain, and trouble themselves as little in the state of one as in the patronage of the other. And here we will not enter into those interesting details, brought forward with so much energy and feeling by the anonymous writer in the Quarterly Review, relative to the patronage of science in the seventeenth century, not merely by the continental sovereigns of that age, but by the court and ministry of Britain; for we should, by condensing, diminish the force and conclusiveness of the argument. A perusal of that statement will show, that among the distinguished philosophers who adorned that age, there is scarcely an individual who did not receive the most substantial rewards for his scientific labours. Newton was appointed successively Warden and Master of the Mint by Charles Montague, afterwards Earl of Halifax, and in the subsequent reign of Queen Anne, "*the then undegraded honour of knighthood*" was conferred upon him. Rœmer in Denmark, Hevelius and Huygens in France, Jacquin and Leibnitz in Germany, the family of the Bernouillis, the celebrated Pallas, and the illustrious Euler in

Russia, and last, though not least, the famous Linnæus in Sweden, are some of the most striking and familiar instances where scientific attainments were rewarded either with high appointments, honorary rank, or liberal pensions; but, what was still more gratifying to the feelings of such men, they enjoyed the confidence of their sovereigns, the converse of their ministers, and the influence they merited. Such an age was not unworthy of that which immediately preceded it, when the sun of patronage arose with such lustre in Italy, and shed a halo of glory over the reign of the Medici, the Emperor Rodolph II., and those sovereigns and princes who courted the acquaintance of Tycho Brahe, and contended for the honour of retaining Descartes at their respective courts. These, and other equally striking proofs of the respect and admiration paid to such men need not be dwelt upon; for it may be urged in explanation, that high scientific attainments were then rare, and were consequently more calculated to excite wonder and respect than they are at present.

(242.) Let us now bring the parallel nearer to our own times; and let us see if, in an age wherein science is more diffused, — and has by this diffusion lost part of its wonderment in the eyes of the multitude, — whether other nations treat it with that indifference and neglect which we complain of. Is France, in the nineteenth century, indifferent to her scientific sons? and does she suffer her philosophers to live unhonoured and unrewarded. Let the names of La Place, Chaptal, Carnot, and Cuvier, created by the government peers of France, and esteemed by the intellectual world as princes of science, exonerate our neighbours from such ingratitude. Wealth, the mammon of *this* country, is *here* considered a necessary requisite for attaining the honours of the peerage: but it is different in all other countries. We know not the average extent of " worldly goods" possessed by the illustrious men just named; but they have never had the reputation of enjoying more than what, among us, would be termed moderate independence. Such, at least, was the case with the Baron Cuvier, the simplicity of whose table and establishment would have been thought mean by a purse-proud shopkeeper of London. Yet, if these and numerous others, scarcely inferior in the republic of science, were not wealthy, they were sufficiently rewarded by appropriate offices in the state, or by pensions, to be placed above the necessity of labouring in matters foreign to science. They were rendered independent, and thus enabled to direct, undisturbed, all the energies of their talents to the respective sciences they have so much adorned. Can we find any parallel instances to these in Britain? can *we* point to such names as Dalton, Ivory, Herschel, Murdock, Henry Bell, Robert Brown, and many others; and say that any one has received honours worthy of such names, or have had the means given to them to secure a respectable independence? Nor is this studied patronage of philosophy confined to France. Turn where we will, either to the leading powers or to the subordinate states of Europe, the same fostering protection shows itself. Prussia has risen to a proud pre-eminence in this respect. The attachment of the reigning family to science and literature is well known; and has spread its vivifying influence to the institutions and the philosophers of that kingdom. Not to mention the celebrated Humboldt, who has been loaded with honorary and pecuniary rewards, and Lichtenstein, the learned and accomplished traveller and naturalist, a long list might be given of other names, celebrated in different departments of science, who repose in the sunshine of royal favour, and are enabled to devote themselves, " in learned leisure," to the investigation of abstract truth. It is a fact well known, that, at " the congress of German naturalists and philosophers, which took place at Berlin in 1828, the attachment of the King and of the royal family of Prussia to the sciences was most strikingly displayed. On the evening of the first day of the meeting, Baron Humboldt, the celebrated traveller, and chamberlain to the King, gave a large *soirée* in the concert-room attached to the theatre. Nearly twelve hundred persons of rank and talent were assembled on this occasion; and the King of Prussia himself honoured this illustrious assembly with his presence. Several princes of foreign states, the Prussian nobility, and the foreign ambassadors, were also present. The princes of the blood mingled with the cultivators of science, and the heir-apparent to the Prussian throne was seen in earnest conversation with the philosophers of his own or other kingdoms that were most celebrated for their talents and their genius."* Science in all its

---

* When Hogarth, indignant at the apathy of our court towards artists and the arts, dedicated his celebrated print of

branches, but more especially natural history, has nowhere received more uniform patronage than in Germany. The splendid works of Jacquin, the celebrated botanist, who travelled for years at the public expense, were given to the world at the cost of the Emperor, and their author rewarded with an office under his patron. No sooner were the enchanting regions of Brazil opened to the researches of European naturalists, than a corps of *savans* were formed at Vienna, completely supplied with every assistance, and conveyed to Rio de Janeiro in ships of war, as a fit retinue to attend the Archduchess of Austria, then united to the King of Portugal. Another fact, by which the comparison we are now making will be better elucidated, regards the celebrated F. Bauer, who, through the strong representations of Sir Joseph Banks, was engaged as botanical painter to the expedition under Captain Flinders. Truly and faithfully did he perform his duties, and returned to England with portfolios filled with inimitable drawings; but our government thought themselves too poor to follow up what they had so well begun: no measures were taken to publish, or turn to any use, what had thus been acquired; Bauer was neglected, and thrown upon his own resources. But this injustice to so distinguished a man was not viewed with indifference by other nations: he was invited

---

the *March to Finchley* to the then king of Prussia, in the same spirit as an English writer would now do, if he selected as his patron the emperor of China, little did he think that, in 1834, the *then* king would be the most munificent patron of art and science in Europe.

to Vienna, with the assurance of protection; and, disgusted with British liberality, he left this country, and ended his days in the enjoyment of a pension from the Austrian government. We are accustomed to look upon Russia as a half-civilised nation, where the arts and sciences are still in their infancy. If this be true (which every year renders more questionable), what is the reason that " so many distinguished individuals of the Academy of Sciences of St. Petersburgh are maintained at the public expense? and that the government has, on all occasions, exhibited the most generous indulgence to her philosophers and artists?" Nor can we hope that a comparison with the minor kingdoms and states of Europe will tell in our favour, or give an indirect sanction to the apathy and ingratitude of England. The court of Bavaria is now, as it was during the reign of the late king, the rendezvous of all men of science and of literature; for *there* they feel assured of being received with the honour due to their high attainments. And here, again, we need only refer to simple facts for a full justification of our sentiments. The possession of the Ægean marbles, now in the Royal Museum at Munich, will for ever record the supineness of Great Britain, in having suffered herself to be outbid for these classical treasures by the little kingdom of Bavaria. The offer was made to both: we declined it, and the Bavarian monarch accepted it. At the time when the scientific expedition to Brazil, above alluded to, was fitted out by the Austrian government, that of Bavaria immediately resolved on following the example, and the king appointed MM. Spix and

Martius, two of the most celebrated naturalists in Europe: they explored Brazil for four years, returned with immense acquisitions, and received from the government not only adequate pensions, but every needful support for publishing their discoveries.* We turn now to Sweden, " which has never been behind the other kingdoms of the North in her zealous patronage of science." We need not refer to Linnæus, honoured, enriched, and ennobled by his sovereign: for that celebrated name belongs to the last century. She yet boasts, however, of her illustrious chemist Berzelius, who has been honoured by a seat in the house of peers, and has been decorated with the cross of the order of Vasa, and the grand cross of the Polar Star; while, in addition to these marks of royal esteem, he enjoys the almost exclusive patronage of the chemical and medical chairs of Sweden. Though circumscribed in its finances, the parliament of Norway has advanced to Professor Hanstein no less than 3000*l*.

---

* In the year 1817, two naturalists, one a Bavarian, the other an Englishman, left Europe, separately, to explore Brazil. They took different routes, and returned to their respective countries three or four years after. The Bavarian published his ornithological discoveries by subscription; and his list, prefixed to the work, contains the names of two emperors, one empress, six kings, one queen, nine princes, four archdukes, seven dukes, and four counts. Total, thirty-four — of the highest titles in the world. (English *none*.) The Englishman, some years after, determined to do the same, and his titled subscribers consisted of one prince (Musignano), and one baron (the Lord Stanley). Total two, English *one*.

to perform his magnetic tour in Siberia. This generous confidence in their countryman has been well repaid by a series of the most valuable observations; and we are sure that every philosopher in Europe is deeply grateful to the patriotic Norwegians for an act of devotion to science, which would do honour to the most powerful nation.*

(243.) Enough, we think, has now been stated, to supply an answer to the question with which we set out; namely,—Is science more neglected by the government of this country than on the Continent?—for we have seen that, with the exception of Britain, Turkey, and perhaps of Spain, "scientific acquirements conduct their possessors to wealth, to honours, to official dignity, and to the favour and friendship of the sovereign."

(244.) We now come to the second question proposed;—Does abstract science, and more particularly natural history, stand in need of any peculiar or national encouragement? We might dismiss this with a very simple reply in the affirmative, by merely asking, whether the pursuit of philosophy will give to its followers the means of living, and will enable them to provide that suitable income for themselves and families, which moderate abilities in other professions will almost always insure? If science will accomplish this, it requires no protection or support; but if, as is notorious, it is the most precarious and the most thankless of all pursuits, encouragement and protection of some sort is absolutely necessary for all those who possess

---

* Quarterly Rev. p. 319.

not pecuniary resources of their own. We speak not here of dilettanti, who amuse themselves—and rationally so — with learning what has long been known, and who, after the ordinary business of the day is finished, make the elegancies of science their recreation. These neither seek or require any other inducement or reward than the self-approbation, and the intellectual pleasure, derived from such a rational source of relaxation; they skim the surface, sip its sweets, but never dive to the depths below. Far otherwise, however, is the case with him who devotes his undivided attention to science in her highest and noblest garb, who consumes days and nights, months and years, in learning all that the accumulated labours of mankind have made known upon his favourite theme, only that he may discover something that they have not; that he may unfold new applications of those general laws already known, trace more clearly the results of their combination, or discover others which open fresh sources of harmony and wonder. The most ordinary mind must immediately perceive, that studies such as these are quite inconsistent with the ordinary business and concerns of life; that they cannot be pursued together; and that, if the depths of science are to be fathomed, and new discoveries brought to light, the task can only be achieved by those whose time is at their own command, whose attention is not divided or distracted by avocations purely worldly, and whose circumstances are such as to make them free from pecuniary cares. Talents, fitting their possessors, for such speculations, must be of a high order, and they are consequently rare; yet still more rare is it to find, superadded to them, the gifts of fortune. From whom, then, if abstract science is to be fostered and rewarded, is this encouragement to come? Certainly not from the public; for what the multitude cannot appreciate, they cannot be expected to reward. If, indeed, the speculations of the philosopher can be turned into immediate advantage by the manufacturer or the merchant, the inventor is in a fair way of dividing profits with the applier; but we are not at present considering such cases. Again, then, let us enquire who are to be the patrons of our philosophy? We live not, unfortunately, in days when any thing of this sort can be looked for from our nobles. "We may in vain search the aristocracy now for philosophers," was the bitter truth extorted from Sir Humphry Davy. If intellectual excellence is so little cultivated among the higher orders, how is it to be expected that they will foster and uphold in others, those qualifications they neither possess nor value? Were it otherwise, we should not see nearly all offices in the state, whose duties implied some acquaintance with science, bestowed upon those who were destitute of such qualifications. Philosophy can, then, only look to national endowments and institutions, or to the favour of the sovereign and his ministers, for that support which she stands in such need of; without which her realms cannot be extended, her discoveries rendered beneficial, or her votaries supported. That this has been the general conviction in all ages, is attested by the uniform agreement of the most enlightened governments to take their philosophers under their own especial protection, and thus to reflect back upon such men a portion of the honour which their discoveries and inventions have cast upon the nation at large. We have now dispassionately enquired into the state of scientific patronage as it is manifested by other nations and by our own; and we have shown that the advancement of all sciences, in reality, stands in need of more efficient encouragement than that which may be expected from the public in general. There are a few considerations, however, which render natural history particularly dependent, for its successful prosecution, upon the assistance and support of national institutions; and these we shall now briefly enumerate.

(245.) Natural history, in the) sense here taken, is restricted to zoology, botany, and mineralogy. And as these branches bear a very unequal influence in their relations to the practical purposes of life, so we must be understood, in the following observations, to allude more to the former than the two latter. Mineralogy, indeed, which forms but a part of chemistry, may almost be considered the only division of natural history which, in an especial and obvious manner, is intimately connected with the wants and elegancies of life. The discovery and extraction of our mineral wealth—the separation and combination of fluids, and the uses to which they are then applied in medicine and in manufactures—at once places mineralogy and chemistry in the rank of the most useful of all the branches of natural history. That discoveries, which eventually have proved extensively applicable to commerce, were never so suspected when their first rudiments were developed, is too notorious to be disputed: for the discovery and the application of a new principle requires very different powers of mind. He who achieved the first, may die in poverty and obscurity; while the other may gain enormous wealth and popular applause. Nevertheless, it is quite obvious that, comparatively to many others, mineralogy is more independent of national patronage for its successful prosecution than either botany or zoology. Not, indeed, that it requires less abstraction of thought, or a less devoted prosecution, but simply on the ground that its knowledge may be turned to practical and pecuniary account. But with botany and zoology the case is for different. Omitting the occasional discovery of a vegetable (like the Peruvian bark), or an animal (like the cochineal), whose qualities prove of universal benefit, a knowledge of these departments can be but rarely and indirectly applied to the ordinary wants of the community; and it is a maxim of the vulgar to esteem every acquirement of this sort, in proportion to the direct benefit it confers on their own interests. Yet because horticulture, which has no other object than animal sense, is thought to be a part of botanical science, the study of plants is more honoured than that of animals, and professorships are instituted for its advancement. Were these more numerous, or were they not strictly confined to members of those universities where they exist, they would, indeed, offer to our veteran botanists the same chance of reward which encourages an adventurer in the lottery, where there is one " capital prize" to about a thousand blanks.

But the zoologist has not even this forlorn hope to look to. Let him spend his youth in travel, his manhood in study, and his fortune in a library and museum, let his labour have been almost as long as his life, he can neither apply the knowledge thus gained to the marketable wants of mankind, or to procuring a respectable competence. Neither can he look, as a last resource, to the hope of some small place of profit, or some slender pension, as a slight acknowledgment from the government or from his sovereign, for that noble disinterestedness which led him to the pursuit of abstract truth, rather than to seek personal aggrandisement in the strife and intrigues of public life.

(246.) Zoology, indeed, may be said to comprehend comparative anatomy, in the same way as mineralogy does chemistry; for both, in fact, regard the analysis of their respective sciences. And it may be urged, perhaps, that anatomy is not altogether in the same deplorable state as zoology We contend not that it is; but we maintain that the two sciences are so vast, that there never yet existed an individual (and we except not one whom the world has just lost) who has reached a pre-eminent station in both. Besides, anatomy, with us, only leads to pecuniary or honorary advantage, when it is confined to the human subject. Anatomical zoology is altogether unproductive of worldly goods; and therefore, with the exception of being followed up, as an amusement, by the wealthy members of the medical profession, it has been long ago resigned into the hands of the pensioned members of the French Institute,—to those, in

EXPENSE OF NATURALISTS' MATERIALS. 359

short, who have no occasion to look around, and see in what manner science will enable them to live.

(247.) While the possession of great zoological attainments leads neither to honorary nor pecuniary advantages, their acquirement is attended with an enormous expense. Books and specimens are the indispensable materials for study. And a large collection of both becomes absolutely essential to every one who aspires to something beyond the minutiæ of his science, the details of names, or the characterising species. Those who are within a convenient distance of the National Museum, are, in a great measure, exempted from such expensive purchases. Yet, when it is remembered that there are no public means of instruction attached either to that establishment, nor to the two leading universities, and that critical examinations, in most instances, can only be made and followed up in the quietude of the study or the library, few will venture to risk their fame on the strength of hasty and partial examinations snatched at a public museum. Natural objects, to be well understood, must be examined and re-examined when the mind is at leisure; when it can discard one conjecture, and, by a fresh inspection, seek to form another: and, if the matter in question has reference to any general law, every animal whose conformation may be thought to bear upon that law, either by affinity or analogy, must, as far as possible, undergo a repeated inspection. The same critical accuracy is necessary in the use of books; wherein a single word will not unfrequently decide a contested point: nor are those works illustrated by figures,—and which, from their high cost, can be possessed but by few naturalists,—less indispensable to our researches. If we aim at great proficiency and superior accuracy, these splendid publications must be had, cost what they will; for few of them, comparatively, can be seen at the public libraries; and the same unavoidable objection exists to their partial and hurried use, in such situations, as that already mentioned regarding specimens. For these reasons, the possession of a library and museum, available at all hours and at all seasons, is indispensable to the philosophic zoologist, who has thus to expend a fortune to become a master in his science.

(248.) But if, after making such sacrifices, both of time and of money, he becomes qualified to write upon the higher departments of his science, to search after general laws, or to unfold a new leaf of the philosophy of nature, and by giving the result to the world, gain at least the praise (unsubstantial though it be) due to his discoveries, his hopes will be miserably disappointed. If he attempt to exhibit his science as a chain of demonstrable truths, and to address his readers as if they already possessed some proficiency in the matter, his work will fall still-born from the press,—no bookseller will incur the risk of publication; well knowing that the little demand for such publications will subject him to a certain loss, even though the work is brought out at the lowest price.* If, on the other hand, he

---

* A striking instance of this has been shown in the *Zoological Researches* of Mr. Thompson, a collection of memoirs in 8vo. published in 3s. 6d. numbers about every three months.

venture to anticipate support from the aristocracy, or the wealthy of the land, by publishing a splendid volume of zoological plates, which, for the beauty of its execution should vie with those published at the national expense of France, his case, even if he succeed in getting royal patronage, is nearly the same. Some few of the court, and still fewer of the nobility, will give him encouragement; but he will be left not a gainer, but a severe sufferer, for his misplaced confidence in the public taste.*
Such are the chief disadvantages attendant on the cultivation of zoology in Britain, and which, independent of the reasons before assigned, give to this science, in particular, especial claims to national encouragement.

(249.) The consequences resulting from this state of things are such as might naturally be expected; and they have been so feelingly described by the able writer in the " Quarterly Review," that we prefer quoting his own words. " Since our scientific men,

---

This work terminated with the fourth number; solely, as then stated, from the want of support to defray its actual expenses! yet of these memoirs it may truly be said, that they surpass in interest every thing that has appeared in this country since the publication of the *Horæ Entomologicæ*. The author makes a direct appeal to the Zoological Society for support, — a society of some 800 or 1000 members; yet, because 150 *subscribers could not be found in the whole kingdom*, these most valuable essays, full of original information, have been discontinued! had the author compiled some trumpery little volume, fit only for the penny press, the sale might probably have reached 5000!

* We here more especially allude to the beautiful folio collection of figures of the *Psittacidæ*, or parrots, by Mr. Lear, a young and most promising zoological draftsman.

then, can find no asylum in our universities, and are utterly abandoned by our government, it may well be asked, What are their occupations? and how are they saved from that poverty and wretchedness which have so often embittered the peace, and broken the spirit, of neglected genius? Some of them squeeze out a miserable sustenance as teachers of elementary mathematics in our military academies, where they submit to mortifications not easily borne by an enlightened mind; more waste their hours in the drudgery of private lecturing; while not a few are torn from the fascination of original research, and compelled to waste their strength in the composition of treatises for periodical works and popular compilations. Nay, so thoroughly is the spirit of science subdued, and so paltry are the honours of successful enquiry, that even well remunerated professors, and others who enjoy a competent independence and sufficient leisure, and are highly fitted by their talents to advance the interests of science, are found devoting themselves to professional authorship, and thus robbing their country of those services of which it stands so much in need." Every one, at all acquainted with the actual state of the physical sciences in Britain, must be well aware that this picture, however humiliating, is not at all exaggerated.

(250.) If we look more especially to zoology, the effect of these discouragements are peculiarly deplorable. So completely have all those higher objects, which entitle the study of nature to the name, and confer on it the dignity, of a *science*. been lost sight of, that there is not one man either in or out of the eight universities of Great Britain, who is at present known to be engaged in any train of philosophic research. The two or, perhaps, three naturalists*, who, during the last fifteen years, have ventured on such classic but now deserted ground, have unfortunately drawn back and relinquished their labours, disgusted and disheartened at the indifference or neglect with which their works have been received. There are few who will bear up against wounded feelings and pecuniary losses, even under the conviction that they are writing for posterity rather than for popularity. Zoology, like all other sciences, is composed of isolated facts and general inferences. If the latter are neglected, there remains only the former: and *these*, being infinitely various, highly curious, and perfectly comprehensible, are thrown into amusing compilations, arranged under some obsolete system, and are then given to the public as specimens of " the science" of natural history. Such is the low tone which this " science" now assumes, merely because no one can be found to act up to the recommendation long since given by the secretary of the Linnæan Society, who points to the absolute necessity that has arisen for generalising the innumerable particulars of which the science of zoology now consists. There requires, indeed, a concurrence of so many circumstances to favour enquiries of such a nature, — talents, time, experience, and independence,— that it is in vain to expect they will be prosecuted, if no sort of encouragement is given to them either by the public or the nation.

---

* M'Leay.   Thompson.   Horsfield.   Annulosa Jav.

That such a reproach should belong more especially to the students of zoology in this country, is but the necessary result of the evils they have to encounter; and we accordingly find the accomplished author, just alluded to, writing as follows:— " English naturalists appear to me, from various causes, to have pursued the nomenclature and examination of species in such a way as very much to exclude from their attention the higher ends of science." In the mean time, cheap compilations are found so profitable, that naturalists, who once shone as original authorities, and are associated with legitimate science, have abandoned such a thankless office; and, drawn away by the lucre of profit, lend their names to speculating booksellers, and assume a station corresponding to the depressed state of British science. Such results, although they may be deplored, are not to be wondered at; "We *have* had," continues Mr. Bicheno, " our Rays, and Listers, and Hunters; but the successful application of our industry has made us a nation of calculators and economists, and, it is to be feared, has almost extinguished that chivalrous spirit inherent to man, that reaps its reward from the honour of the cause in which it is engaged. " The criterion of value of every thing in England," continues this able writer, " is" its marketable price; until the highest stations in the state, and the most honourable conditions of inferior society are measured by their stipend."* It is therefore too much, perhaps, for us to expect, that, in the present state

---

* Address delivered at the Zoological Club of the Linn. Society. 1826. By J. E. Bicheno, Esq. Sec. L. S.

of things, British naturalists should toil and labour in the higher branches of their science, only to receive mortification, when, with so little trouble, they can enrich themselves, and insure popular applause, by working up the materials of others.

(251.) Before we proceed further, a recapitulations of the facts already stated,— in reference to the present state of nearly all the physical sciences, and particularly to that of zoology, — will not be misplaced. If, as we have shown, admittance can be gained by purchase into all our learned societies; if there are no national institutions, whose officers are selected from among the ranks of science; if there are no honorary distinctions, as in other countries, peculiarly appropriated to our philosophers and men of letters; if no pecuniary rewards, or retired pensions, are bestowed upon those who, above all other ranks, have mainly contributed to the true glory of the empire; if the physical sciences form no part of the system of education taught at our universities; if there are no professorships, or no means of instruction for aiding and encouraging the study of the material creation; if works on abstract science entail loss upon their authors; if, in short, these are things " which be," can it excite surprise for a moment, that the taste and the possession of true legitimate science has declined in Britain, while it has advanced on the Continent? Can it be wondered at, that those, whose love for abstract truth are leading them still to pursue it, despite of neglect, mortification, and discouragement, should remonstrate plainly and perhaps indignantly against such a state of things? Can it be said that eminence in science should alone be excluded from those honours of the state which are liberally bestowed upon " the mere possessors of animal courage," or the enjoyers of mere wealth? Surely there must be something in all this, quite inconsistent with the national character for liberality and right feeling. Men of the highest talent, pursuing different studies, and therefore viewing the same question from different bearings, belonging to no political party, upbraiding no particular ministry, and having no selfish interests to warp their judgments, — such men, we may safely conclude, would not simultaneously raise their voices, and proclaim their own dishonoured state, without urgent cause. Let us at least be candid, and fairly admit there must be something which calls for amendment; and under this spirit, those who possess the power will be better qualified to consider, in coolness of judgment, what measures are best calculated to restore the science of Britain to a healthy state.

CHAP. III.

ON THE MEANS POSSESSED BY THE GOVERNMENT AND UNIVERSITIES FOR PROTECTING AND ENCOURAGING SCIENCE.— ON TITULAR HONOURS.

(252.) In extending the foregoing reflections to the suggestion of means for obviating the evils therein complained of, and for giving to the science of the country that efficient support which it so much requires, we feel that we are entering upon a subject of difficulty and delicacy. Those who are averse to the innovation of established customs, institutions, or modes of thinking, are always more numerous than those who imagine they can be improved. This feeling is natural to the mass of mankind. Few have either the energy, or the inclination, to look deeply into things which they have been accustomed to see go on, year after year, in the same course; and which, they therefore conclude, require neither alteration nor amendment. Say what we will, the mind leans with a degree of fondness, if not of veneration, to every thing which has the authority of antiquity, or of long-continued usage; and these feelings are increased, if those whom we most esteem, and who may have to administer our ancient laws, conscientiously defend their continuance. On the other hand it is to be remembered, that all institutions, to be extensively beneficial, must be altered and modified to suit that progressive improvement which is the consequence of good government. So plain a truth as this, none can be found to deny in the abstract; but the moment we come to apply it in its particulars, — to single out any one case which, for assigned reasons, would appear to warrant timely but effectual amendment, — our prejudice against innovation returns with its former force; we either forget the general assent to the axiom, that all human institutions should be adapted to the national state of civilisation, or we are prone to contend, that although moderate reformation is in the main beneficial, yet, in the particular case pointed out, it is uncalled for, and therefore unnecessary. But the ingenuous mind, anxious to discover truth, will not suffer predilections to turn it aside: it will calmly and patiently investigate arguments opposed to its own impressions; it will concede such points as appear supported by sufficient evidence, and if, on mature reflection, it rejects others, it will give to its opponent the credit at least of being actuated by a pure and honest spirit of dissent against the thing complained of. Where these feelings are mutual, controversy, in all matters, will be denuded of those baser passions with which human infirmity has clothed it. Truth, unchanging truth, would be the only object sought, and an honest and a good mind will receive almost equal satisfaction, if the treasure is found by another, rather than by himself. It is to such minds, and such only, that we now appeal. For, however warmly we may feel,

PLAN OF THE ARGUMENT.

and perhaps write, upon a subject so dear to us as zoological science, we wish to express our sentiments with all the moderation that can be consistent with a strenuous defence of opinions. We are sensible that all these may not be correct, but it is hard to believe that some are not fully borne out by the evidence produced. For those who may conscientiously differ from us, we hope to preserve the same good-will as heretofore, and we only ask the same from them. One thing may be safely said, that, in discussing the state of science, and becoming a humble suitor in its behalf to those exalted few who have the power of honouring its professors, no feeling of a personal nature is to be answered. We can plead the cause of others, and rejoice in their honours, although physical incapacity and confirmed habits of seclusion will ever prevent individual participation. If but one firm and efficient step is taken, by those in power, towards reinstating the science of Britain in that pre-eminence she once held, we shall be amply repaid for the irksomeness of conveying censure, and of criticising public institutions.

(253.) In discussing the question now before us, we shall endeavour to point out the most effectual means by which zoological science may be promoted and upheld; first, by the universities, and secondly, by the government.

(254.) If tastes are to be formed, and feelings implanted in the human mind, they will never so firmly take root as in the spring of life. Hence, if we may indulge a hope, that science may eventually hold that station among us which it elsewhere enjoys, we must look, mentally, not to the present, but to a succeeding generation. To expect that such changes, as great as they are necessary, should take place in the minds of those who have long been satisfied with the present state of our learned and scientific institutions, would be chimerical. Some little advance, indeed, may be hoped for, because a few steps have already been taken; but Time, who works slowly, but surely, is the best reformer. The conviction of truth is rarely, if ever, sudden, while violent changes, besides being in opposition to the whole analogy of nature, have the acknowledged evil of generally destroying that which only required renovation. Where popular clamour, also, is vehemently raised against any particular establishment, there is danger that its true friends, by suggesting amendments, may be confounded, or at least be thought indirectly to co-operate, with its enemies. We seek to amplify and adorn with new pinnacles, neither to hide, much less to level with the ground, the olden towers and spires, the columns and the domes of our collegiate and time-honoured structures. Nevertheless, we feel, upon so important a subject as the present, the impossibility of avoiding all allusion to our universities. These institutions have long been considered—and in most respects justly—the seats of British learning. They are alike venerable for their antiquity, the noble feelings which led to their foundation, and the bright and hallowed names with which they are associated. They were founded for the education of those higher ranks of the empire whose feelings and conduct were to give a tone and an example to the great mass of the people; and that this was the *spirit* in which they were first instituted cannot for a moment be doubted, a fact which should be ever borne in mind by their governors, and which authorises such deviations from the strict letter of their laws as the altered circumstances of the times may require. Let us not, however, upon so important a question be misunderstood. All the branches of university education arrange themselves under two distinct heads. The infusion of the national religious creed, and the study of ancient literature,— the expounding of the book of God, and the study of the works of man. On the first and greatest of these objects time can have no effect. The Holy Scriptures are the same to-day as they were eighteen centuries ago; nothing has been added to them, nothing has been taken away. These holy bulwarks of our faith are unchanging and unchanged; and they require studying with the same earnestness and the same devotedness now, as when these venerable sanctuaries of the church were first founded. But, in regard to *human* wisdom, the case is different; the last century has witnessed surprising changes not only in the progress, but in the kind of knowledge necessary or desirable to be taught. Sciences, which were scarcely known by name to the founders of our colleges, have assumed form, extension, and demonstration; while others, utterly unknown to our ancestors, have started into life, and, like the overflowings of the Nile, have spread over the land, fertilised its provinces, and are now producing, in an infinity of ways, a fruitful harvest of palpable good to its inhabitants. It is to these sciences, and to these only, that our present observations relate. Who, then, that reflects upon the original intentions by which the founders of our universities were actuated, — who will maintain that the education of the higher classes is to be confined to the same studies in the nineteenth century, as were taught in the seventeenth? that many of the most intellectual, as well as the most elegant branches of physical science should be excluded from the regular course of university education, or, if they are permitted to be taught, that the option of learning them should be left to the pleasure of the students themselves, without any enforcement arising from the rules of their college,—any inducement held out to stimulate their exertion, or any to reward their acquirement? The usual reply to these interrogations is, that an acquaintance with the physical sciences formed no part of the original institution of our universities, and, therefore, to introduce them as secondary objects of study would be in direct defiance of their charter. But this objection has been already anticipated; and if another answer is required, it may be found in the close connection between *natural* religion, which is so strongly elucidated by the physical sciences, and that *revealed* religion, which it is the business of our universities to uphold and expound.

(255.) This connection of the twofold causes of our homage to the Great Creator, has been so admirably illustrated by the talent and eloquence of one of the brightest ornaments to science now among us, that it would be useless to repeat the same arguments, and hopeless to place them in a stronger point of view.* If, then, the material world is replete with proofs, innumerable and unanswerable, not only of the being of a God, but of His infinite power, wisdom, and bounty; and if, above all, these temporal things speak to us, as in parables, of those eternal destinies with which man is inseparably linked, the study of the visible creation is *second* only in importance to that of the spiritual. Ancient literature, whatever may be its advantages, however it may, *judiciously selected*, refine the taste, improve the diction, or inform the understanding, cannot for a moment be brought into equal comparison with the sublimity, the pureness, and the exalting nature of natural history. The student of the one draws his materials of thought from the works and deeds of man; the other studies from the interminable library of nature, and from the examples so brought before him, learns to exercise towards his fellows, however imperfectly, that beneficence and compassion, and that unwearied solicitude which he sees is extended by his Maker to the meanest insect that crosses his path. The Newtonian philosophy, indeed, expands the mind to such a painful degree as to make it fall back upon itself, as conscious of its inability to grasp the full range of the sublime truths it dimly unfolds, or even of the effects which those truths produce in the visible creation. Yet the wonders of the heavens, however awfully magnificent, and

---

* A Discourse on the Studies of the University of Cambridge, by Adam Sedgwick, M. A. Woodwardean Professor.

speaking a language peculiar to themselves, are yet denuded of those circumstantial details which are more suited to our limited faculties, and in which the generality of mankind can not only feel an interest, but a pleasure.

(256.) We may liken these different emotions to those entertained by a traveller, who from some distant eminence first gains a view of Mount Etna, dilated into its full dimensions, its long extended outline unobstructed by a single object, rising gradually from the watery horizon on one side, and from that of vast plains on the other, until its pointed summit seems to touch the firmament. Here and there a deep line of ravines may be traced, and darker stripes indicate either regions of forests or extinguished rivers of lava; but beyond these obscure appearances nothing can be made out on the sides of this mighty mass, and which seems to circumscribe half the horizon. The eye of the traveller, indeed, seeks not for details: his mind is absorbed with ideas of vastness and indefinite sublimity: there is no room for lesser feelings, and there is no power of gratifying them.* But, when he descends from his station into the plains below, and after two days' travel begins to ascend the sides of that stupendous mountain, whose details he is now to

---

* I have here attempted to describe, most inadequately, the view of Mount Etna which bursts upon the traveller from the heights of Taormina; but no language can do it justice. I have never seen, either in the old or the new world, a prospect of such transcendant magnificence. See Denon's Sicily, p. 15., and Brydone's Tour.

explore, a new set of emotions, totally unlike those he before experienced, arise in his mind. In his long and gradual approach, the mountain itself seems to have changed its shape and its character: instead of one sublime and simple whole, it appears to have separated itself into innumerable ridges of gradual slopes, abrupt cones, or frightful precipices; these again, as he advances further, seem to contract themselves more into ordinary dimensions, until, but for an occasional opening, from whence the apparently sunken summit peeps forth, he might fancy he was merely traversing a hilly or mountainous country. His pleasurable feelings *alone* are now excited, as he passes through the little villages, talks with the people, gathers a plant, catches an insect, or picks up a mineral. He enters, in short, into details which he can understand. He can now examine and explore what he sees; he is busied with things more suited to his every-day powers of contemplation, and, if his thoughts do not rest on the purely sublime, they are not pained by being overstretched.

(257.) Such may not be thought an unapt illustration of the different effects produced upon us by the respective studies of astronomy and zoology: both have immediate reference to the power and wisdom of God; but the one is more suited to the generality of mankind than the other, and brings His attributes more home to their understandings. This quality being granted, does it not follow that it should be encouraged, fostered, and protected, as the most appropriate adjunct to revealed religion of all the physical sciences? Should it not, in fact, be inseparably connected, in every system of education, with the study of spiritual truths. Natural history is the most appropriate handmaiden revealed religion can receive; she is always at our side, ready to point out in every plant that grows and in every creature that breathes, the verity of those things which are unseen; things which the youthful mind, however unaccustomed to reflect, is nevertheless instructed to believe in. But there is danger, it may be said, in two ways, in thus making zoological science one of the essentials of an academic education. Firstly, that as the science in its present state exhibits none of those philosophic generalisations and definite laws to be found in the astronomic world, the mind may become too much attached to its minute details, to dwell upon the lessons or inferences they should teach; and secondly, that as natural history is rather a contemplative study, its acquirement would involve more time than can be spared from studies more immediately bearing on the active duties of life. Both these objections, more especially the latter, appear good, and therefore deserve our serious attention.

(258.) No fact can speak more plainly of the consequences resulting from the disregard of zoological science in Britain, than that it is the only one in which (until very recently) no general laws had been discovered. Other branches of physical science have had their Keplers, their Newtons, and their Davys, who have each, by slow but unwearied inductions, reduced a multiplicity of appearances to a few lofty generalisations, under which an innumerable diversity of facts, formerly isolated and apparently anomalous, are brought together, and are shown to be but modifications of one and the same principle. Natural history alone has hitherto remained unhonoured by such names. Linnæus saw the station which his favourite science should hold, but, with so few materials and facts before him, he wisely abstained from attempting philosophic generalisations. Much, indeed, has been said, by those who should have known better, about what has been termed the *law of co-relation*, of which a late celebrated naturalist of France has been extolled as the discoverer, but which has been known to every naturalist since the days of Aristotle; this law of co-relation being, in fact, no other than that the structure of an animal is adapted to its economy and habits.

(259.) Such being the state of the philosophy of zoology, can we imagine, that if its cultivation had been fostered, it would not have reached a higher altitude in the rank of the demonstrative sciences? Are we to suppose, for a moment, that it is exempt from the influence of definite laws, and that the almost infinite variety of form and structure in the objects it embraces, cannot be reduced to a few primary types, or that the mode of their variation is fluctuating and indefinite? If we reject all such suppositions, as being at variance with the whole analogy of nature, under what circumstances can we suppose such discoveries are most likely to be made? Certainly by those whose minds have been disciplined in the universities, and who have not only acquired a love for *abstract* truth, but who are qualified to pursue it in a philosophic spirit. Now, if the very elements of zoology cannot be acquired at these seats of learning, how can it be supposed that a taste for it should be acquired in youth and cultivated in manhood, when the student emerges from college, quits education, and at once enters upon the active duties of life? But supposing that professorships were appointed, on small salaries, sufficient, with the emoluments of lectures and pupils, to make them desirable. What leisure is left for the lecturer to prosecute original research? The emoluments of his chair being chiefly derived from teaching the elements of science, these will naturally engross his chief solicitude. He will strive to make his lectures popular, by waiving the discussion of abstract principles, and dilating on all those comparatively trivial matters, which his audience can at once understand: under such circumstances, how can he himself himself cultivate or teach to others the higher principles of science? or how can he concentrate his mind to the exclusive study of one or two abstract theories, which, after occupying his deepest attention for years, may be expressed in a few lines? It is from among men of talent, and of "learned leisure," who from their station in society possess competency, that we may hope zoological science will be pursued with true dignity; from such only may we expect its *advancement* instead of its *diffusion*. Characters, promising to unite all these necessary acquirements, are most likely to be formed at our universities; and if no effectual means are supplied for directing such powers where they exist, what wonder is it that zoology is looked upon as a mere vocabulary of technicalities, or an amusing volume of animal biography? We admit, therefore, that in the present state of things, the mind of a young zoologist would be chiefly occupied in the minutiæ of his study, because, as to the science itself, he would have little else to reflect upon; he has neither been instructed, nor has he heard of definite laws, in zoology, although he may be acquainted with those of the Newtonian philosophy. But it by no means follows, that because he busies himself with minute details, his mind will therefore receive a corresponding contraction. The least, no less than the greatest of the Creator's works, possess the power of exciting the loftiest ideas of his power and wisdom, and while emotions such as these, fostered by previous studies, rise in the youthful breast, it is of little consequence to his happiness, whether he is engaged in the investigation of general laws, or examines, under the microscope, the complicated structure of the mouth of an insect. To stigmatise such pursuits, carried on with such feelings, as trivial or mean, is not only folly, but gross impiety.

(260.) But natural history, it may be said, being as much an intellectual as a spiritual study, cannot well be made a part of university education; inasmuch as its acquirement would trench upon time absolutely requisite for other studies more immediately bearing on the active duties of life. Now, if our universities were schools for commerce, manufactories, or the practical arts, where young men were to be instructed in those professions by which the great machine of active society is carried on, this objection might have some weight. But a moment's consideration on the particular classes educated at college will show the weakness of such reasoning: they consist almost exclusively of the sons of the nobility, of the clergy, and of the wealthy gentry: the first and the last are almost exempt from the necessity of following *any* profession, while those who are intended for the church are equally freed from the obligation of acquiring a theoretic knowledge of their calling (as the physician and the lawyer are obliged to do), *after* they have quitted the walls of the university. Academic studies are rather intellectual than practical; that is to say, they have no fellowship with the commercial and manufacturing — the military, the naval, or the empirical arts: how important is it, then, to the future happiness of young men, educated and nurtured with such feelings, to infuse into their minds a love of physical science, — to supply them with intellectual and pure resources in after life, suited at once to those habits of abstract reflection they have acquired at college, and to the leisure which attends upon rank and wealth? Some, indeed, will be called into active life, and will be destined to fill important stations in the state; but these, in comparison to the majority, are few; and if our statesmen and legislators had been early impressed with the beauties or imbued with a taste only for philosophy, and had been better instructed in its objects, the science of Britain would not, at this time, be so utterly neglected.

(261.) But by far the larger portion of those young collegians, not destined for the church, on finishing their education, enter upon a life of indolence and pleasure, without having imbibed a taste for any one philosophic or intellectual pursuit, which might be followed, if not as a study, at least as an elegant amusement, suited to the education they have received, the advantages they enjoy, and the superior station of society in which they move. How frequently do we see young men of naturally superior abilities, after gaining university honours, to which they were excited by the short-lived stimulus of competition, leave their college, and settle down upon their paternal estates as mere country gentlemen, hunting squires, or racing patrons: filling situations, in short, which should be occupied by men of a lower grade in the scale of intellect, but which they fall into, merely because they have not been instructed in any pursuit which will call into continued activity those abstract powers of reasoning or observation they may have acquired at college. The mind, however high may be its natural capabilities, invariably sinks to a level with its usual occupations. And science, being neglected by these men, who have almost exclusively the power of pursuing it with true dignity, is left to those who are obliged in most cases to connect it with objects of trade, or of pecuniary advantage.

(262.) But if a taste for natural history is so well calculated to give elegance and dignity to the recreations of the aristocracy, how much more is it in unison with those feelings and habits of thought which should belong to the young clergyman when he quits his college, and desires to enter upon the sacred duties of his profession? The excitements of collegiate studies are now over; competition is at an end, and he either waits to be called to active duties, or he enters at once upon his divine ministry. In the former case there is leisure, more than enough, to make himself practically conversant with the wonders of the material world; for into whatever department of natural history his inclination may lead him, our island possesses stores of objects calling for observation and research. Whether his attention be directed to zoology, botany, mineralogy, or geology, he cannot fail of drawing from one and from all these studies, materials for illustrating the perfections of Him whose word he is to teach, and whose works he is to "magnify in the congregation." If, on the contrary, the young divine is settled on a small curacy "remote from cities," what a never-failing resource would he find in prosecuting those physical studies, the elements of which had been acquired at the university. Independently of the spiritual use to which he could apply material things,—the shadows of such as are heavenly,— how little would he feel the loss of ordinary society, and how little would he prize that which usually distinguishes country families. It is for these reasons we contend that the interests of religion, and the future worldly happiness of the students, are most materially concerned in the present question regarding our universities. It may be said that these pursuits may produce evil, by absorbing too much of the time and attention of the young clergyman from his pastoral duties. But this is no argument to the purpose, for all the good and all the virtues of this world might equally be prohibited for the self-same reason.

(263.) The neglect of this science at our universities is naturally followed by a similar neglect on the part of government. That government being carried on, for the most part, by those who, as youths, saw how slightly such acquirements were held at college, and who now, as men, look upon them with the same feelings of indifference. The one is the natural consequence of the other. Until those to whom the executive government of a nation is intrusted, are impressed with a respect, if not a love, for philosophical excellence, and are fully convinced of the important influence it exercises over the welfare of the community — in an infinity of ways, — it is obviously hopeless to expect more than a partial, if not a merely nominal improvement. Nevertheless, *any* encouragement in the present state of science, emanating from the government, may be regarded as beneficial, although it may fall very short of that which is necessary, or which exists in other countries. It is encouraging, therefore, to observe that, under the administration of Lord Goderich, the sum of 1000*l*. was devoted to the expenses of bringing out the zoological discoveries of the arctic expedition; and that another sum, in like manner, was appropriated to the publication of Captain Beechey's acquisitions. Yet so slight is the estimation in which such publications are viewed by the public at large, that, even with these helps, the publishers have made no scruple of complaining bitterly of the pecuniary loss that has fallen upon them; so that unless, on some future occasion, the government can be persuaded to contribute more largely to the publication of discoveries made at the *national* expense, it may be fairly questioned, whether any respectable bookseller will undertake the risk of publishing them. The French are well aware of the necessity of national patronage to such works; and with that munificent liberality which characterises all their proceedings regarding science, they annually set aside a considerable sum for this exclusive purpose. It has been solely owing to this liberal spirit, worthy of a great nation, that the splendid zoological discoveries made in the voyages of Dupery and other navigators have been published in a style of beauty and completeness, which is no less valuable to science than honourable to the nation.

(264.) But let us proceed regularly; and give to each of those means by which an enlightened administration can encourage science, a separate consideration. These appear to be as follows: — 1. The appointment of scientific men to those offices wherein their acquirements can be made subservient to the public good. 2. By aiding and assisting our universities in the establishment of professorships. 3. By condescending to consult, on all such questions as relate to science, those scientific institutions of the country which are the proper tribunals for deciding such questions. 4. By removing all those impediments and regulations which press upon the authors of illustrative works on natural history, and for the general encouragement of such works. 5. By bestowing honorary or pecuniary rewards upon those whose discoveries or researches reflect honour upon the nation.

(265.) I. The first subject which demands the attention of any administration desirous of placing the science of this country on a firm and prosperous basis, relates to our universities, and to the establishment of Regius Professorships in those branches of physical science wherein there is at present no sort of instruction. The important advantages which would result from this measure have already been sufficiently dwelt upon. At present, it seems to be the general opinion, that our government (so far from incurring any expense by the annual grant of 2000*l.* made by the nation for " defraying the salaries of professors in the universities of Oxford and Cambridge") receives the sums back again into the treasury in the shape of taxes of the most odious description, inasmuch as they are extorted, not indeed by a sale of honours, but by taxing those who achieve honours; those, in fact, who proceed with credit through an expensive term of years of collegiate discipline, and are therefore admitted to their degrees, so that these fees become a direct *tax upon learning**, in those very establishments avowedly made for its encouragement! What is given by the government with one hand, is taken back, with more than usurious interest, by the other. The peculiarly oppressive nature of these exactions have recently called forth much public complaint, and they are altogether so opposed to the spirit of the age, that we think the subject only requires to be mentioned in parliament, to give our present ministers an op-

---

* Every Bachelor of Arts pays to the government *three,* and every Master of Arts *six* guineas on taking his degrees. Different sums, varying in amount, are imposed on other degrees. See Appendix.

portunity of at once removing such a stigma upon the government. The salaries, moreover, thus voted to the professors, at one at least, if not at both, of our universities are, with two exceptions, so small, as to be quite paltry.* It is probable, however, that there may be local or unexplained reasons for this inequality with which I am unacquainted.

(266.) It will be seen from a letter by the Rev. T. Newcome, M.A., printed in the Appendix, that the average amount of this tax upon learning may be fairly estimated at 5000*l.* annually: while the sum allowed by the nation to the professors is only 2000*l.*, leaving a profit to the government of 3000*l.* a year, drawn from the purse of parents, generally not rich, and mostly possessed only of life incomes, and of those whose industry and talent alone support the reputation of our universities. Now, all that is required from the government is the transfer of the proceeds of this tax into the university chest, for the purpose of paying their own professors, and managing their own concerns. The

---

* According to the Cambridge Calendar they are as follows: —

	£
Regius Professor of Divinity	40
——————— Civil Law	40
——————— Physic	40
——————— Hebrew	40
——————— Greek	40
Professor of Chemistry	100
————— Modern History	400
————— Botany	200
————— Mineralogy	100

tax is certainly oppressive upon those who least merit such a "reward." But even allowing it to remain as it now is, the character it would assume, when its proceeds were applied to the liberal endowment of professorships, would be quite different. It would be like a private subscription, raised among friends, for the purpose of making up a purse for the benefit of some one or more of their own number: each member would gladly contribute his share, under the conviction that the prizes will be awarded to the most deserving; and that, with proper exertion, he has as fair a chance as any other aspirant of receiving back his fee, augmented an hundred-fold. Were it customary to choose the members of our universities more from the ranks of science, than from those of politics or of arms, this grievance probably would not have so long continued in operation. At all events, it appears so easy of redress, and will so effectually remove all further complaint against the government in connection with the state of science at our universities, that we trust in having the power of omitting all such censure in another edition. There can be no doubt that the heads of our universities will most joyfully accept of such a boon, and be glad to be thus rid of an appearance only of receiving government bounty, while it cannot be doubted that they will use it freely and effectually for the promotion of that philosophy which is properly connected with Christianity. Every reflecting mind must participate with such authorities, in their fear and dread of hasty and inconsiderate changes, or of that "radical reform," now so loudly called for. The whole analogy of nature shows that gradual progression, or slow developement, is one of the primary laws both of the moral and the physical world.

(267.) Nothing gives so much stability to a government, or operates more beneficially to the welfare of a nation, than a judicious selection of its executive officers, that is, of singling out, from among the great mass of the people, those individuals who, from possessing talents or qualifications of a particular order are best qualified for performing certain duties. An axiom so plain and acknowledged would scarcely require notice, was it not with us so repeatedly violated for the sake of strengthening political party, or of rewarding political services. Without expecting that any of our philosophers should be cabinet ministers, or privy counsellors, or ambassadors, although such a union may be found in Continental courts, it might reasonably have been supposed that in a country like Great Britain, the extent and variety of her public institutions would have furnished ample provision for scientific men. As mistress of the ocean, her Board of Longitude should, like that of France, have furnished an appropriate endowment for her men of science; her lighthouse Boards, with their immense revenues, might, like the corresponding Board in France, have supplied situations to others; her "Boards of Trade" might have been appropriately conducted by men who combine practical with theoretical knowledge; her mineral treasures might have proffered a tithe of their produce to reward the knowledge which explored them, and applied them to the arts; her Royal Societies might have added several official situations; and her universities, besides the ordinary chairs for professional education, might have contained others, which, while they attracted men of great name within their precincts, left them sufficient leisure to pursue their researches. All this might have been expected in England, for this simple reason, because it is found in other countries, less able and less called upon, to be liberal to their philosophers. But how stands the case with us? The Board of Longitude became almost useless, from being occupied by unscientific men; it was thus brought into disrepute, and was abolished in 1828. Of the three lighthouse Boards, " by that fatality which impends over every British institution," not one of all the numerous members and officers is a man of science, or is even acquainted with those branches of optics which regulate the condensation and distribution of that element which it is their sole business to diffuse over the deep. That boards so constituted are totally disqualified to judge of improvements or inventions offered for their adoption, is quite natural. There is a remarkable instance recorded by the writer we have just quoted*, wherein the inventor of a new compound lens, after vainly endeavouring to draw the attention of our boards to his discovery, had the mortification of seeing it claimed some years after by a learned foreigner, and universally introduced on the coasts of France as a new and important improvement in lighthouse illumination, while the lighthouses upon our shores, proverbially

---

* Quarterly Review.

the most dangerous in Europe, are still, as it is asserted, illuminated by the " old unscientific methods." If any misapplication of patronage calls for an immediate reform, it is surely in these establishments, wherein is largely involved the risk of human life, and the loss of large property. Mr. Babbage adduces another instance, bearing strongly upon this exclusion of scientific men from the national councils, which is even more to the purpose. " To those who measure the question of the national encouragement of science by its value in pounds, shillings, and pence, I will here state," observes Mr. Babbage, " the following fact:—A short time since it was discovered by government that the terms on which annuities had been granted by them were erroneous, and new tables were introduced by act of parliament. It was stated at the time that the erroneous tables had caused a *loss to the country of between* 2,000,000*l. and* 3,000,000*l. sterling.* The fact of the sale of those annuities being a losing concern, was long known to many, and the government appear to have been the last to be informed on the subject." Now it is perfectly clear, that if the government had condescended to consult our mathematicians, before they legislated on matters they were confessedly ignorant upon, this enormous loss to the public purse would not have happened. Had one half of this sum been judiciously applied to the protection and encouragement of science, institutions might have been endowed, professorships established, and pensions provided for our philosophers, which would have placed the scientific establishments of the nation on an equality with those of all others.

(268.) It were needless, after this, to enforce, by other instances, the benefits which would follow the appointment of scientific men to situations under the government, wherein their acquirements might be of service to the state, and their opinions taken upon all such questions as came within their respective provinces of science or of art. Nor are the pursuits of the naturalist altogether devoid of public utility. On questions regarding the fisheries, the capabilities of our colonies in supplying new articles of commerce from their natural productions, and many others touched upon in another part of this volume, our legislators need not be ashamed of seeking advice, or at least information, on matters upon which it cannot be expected they should be competent judges.

(269.) But it is not only in the home departments that Great Britain possesses ample and appropriate means of making honourable provision for men of knowledge, while she receives in return the benefit of their services. Her extensive range of colonies, with their numerous establishments, and her consulships—spread over every part of the world—afford situations peculiarly well adapted for those enterprising spirits who ardently desire to study nature in other climes, and under other aspects, than those at home. Whatever may be said to the contrary, we know, from personal experience, that the duties of colonial governments, however necessary, are generally very *light*, admitting of much leisure, which an active and powerful mind would turn to good account. Our consulships, with few exceptions, more especially, are of this character. And from the little previous knowledge requisite to discharge their duties, they would be peculiarly appropriate and acceptable to young men of science, already sufficiently acquainted with mercantile concerns, or perhaps still engaged therein, but who desire some little leisure for more intellectual though profitless pursuits. There are, for instance, five or six consulships, at the least, in South America, where the duties seldom occupy more than three or four hours in the day, although the appointments are indispensable. Why cannot we follow the example of France, in this instance, at least, and give such situations to those who (being duly qualified in other respects), are men of science, desiring to visit other countries, and who, in return, would enrich our national collections with new objects, and our scientific transactions with fresh discoveries? The supineness of our government on this subject was particularly remarked by some intelligent foreigners a few years ago, when the Brazilian Consuls of Russia, Prussia, and France, at Rio de Janeiro, were all naturalists, having full leisure to perform their official duties, and at the same time to collect and transmit to their governments large and valuable collections of Brazilian zoology. The English Consul, at one of these ports, on the other hand, was an illiterate person, who turned into his hammock, and dozed and smoked away the greatest part of that time which his official brethren were so beneficially employing. We remember that the British consul at Athens in 1812 was a Greek, and we found one of the vice-consulships in Sicily given to a Frenchman.

Surely there might have been found among the young independent students of nature, in this country, some who would have accepted these situations with small salaries and less work, for the sake of the leisure they allowed, and the respectability they gave.

(270.) There is yet another and a very important mode by which a liberal government can provide both for the advancement of science and the employment of her votaries: we mean the appointment of scientific men, of known reputation, to accompany our voyages of discovery. These opportunities, indeed, are " few and far between;" a reason which might be urged as the best for conducting them with a liberal spirit in their minor details. But here, again, we have reaped neither honour nor credit. Not to compare the French expedition to Egypt with our own,—the one accompanied by a splendid train of the most eminent savans of France, the other without a single philosopher,— we need only think on the different fates that attended two of the more celebrated Egyptian travellers, Denon and Belzoni; the one, honoured, patronised, and enriched by the favour of his government; the other neglected, dishonoured, and heart-broken. Belzoni, impoverishing himself to accomplish that which the British government should have felt honour in patronising, and thus leaving his widow dependent on the casual bounty of strangers. Where, again, are the zoological or botanical results of Flinders's voyage? where those of the Congo expedition? or those of Ross and Parry? We admit, and we do so with pleasure, that a better spirit has begun to show itself; and that, during the enlightened administration of Lord Goderich, Europe, for the first time, beheld the government of this country giving to the world the fruits of her zoological discoveries. But to do this effectually, a more liberal system must be adopted in the outset. Our present method (certainly a very economic one) is to impose the duties of "naturalist" upon some one of the regular officers of the vessel; at a time when such duties, to be effectually performed, would require the undivided attention of three or even four persons, each taking different departments, and all requiring the assistance of a professional artist. It is not enough that the naturalists of such expeditions should bring home stuffed skins, dried bones, or things in spirits; for such duties can be done by the sailors themselves. It is not a knowledge of dead skins, but of living creatures, that we now want. It is an acquaintance with the internal structure of animals, and the true shape of the soft *Mollusca* (which latter are lost or distorted the moment they are plunged into spirits), that will effectually advance zoological science. Now all this information can only be acquired by studying and drawing the animals when fresh; and it therefore follows, that such researches and duties can only be made and done by experienced naturalists and professional draftsmen. Here it is that the government of France, from possessing scientific advisers, has gained the highest credit, and has done more to advance the knowledge of invertebrated animals than any other nation whatsoever. All their expeditions have been accompanied by experienced zoologists, botanists, and draftsmen, each taking some specific department, and prosecuting their researches on recent subjects. These new acquisitions are then published at *the sole expense of the government*, and in a style of magnificence worthy of a great nation. We need no other proofs of the talent thus called into action, or of the liberality which fostered it, than the splendid and invaluable series of zoological folios containing the discoveries of Peron, Quoy, and Garnot, Lesson, and those of the naturalists and artists which accompanied the Astrolobe discovery ship, now in course of publication.

(271.) It follows, from necessity, that if men of science are once allowed, like other ranks in society, to aspire to the honours of the state, by the patronage and protection of " the fountain of honour," the government of the country possess, in them, the best advisers, and the purest means of information on all scientific questions, that can be found. But we must *appreciate* excellence, from a conviction of its worth, before we condescend to ask advice of others, and before we can be persuaded that we ourselves are incompetent judges. So long as the influence of scientific knowledge upon the business of life is neither perceived nor valued, so long will its services be neglected, if not despised. On the other hand, when once this connection is seen and acknowledged, our philosophers will be looked upon as fit advisers on all occasions wherein their acquirements bear upon the question at issue. We are again compelled to cite the institutions of other countries as patterns for our own. The members of the French Institute*. appointed and paid by the government, became, in return, its scientific advisers. There is no establishment of this nature in Britain; but if the Royal Society was placed upon a better footing, many of the advantages derived by the French government from its Institute might reasonably be expected from the oldest, if not the best as they are at present constructed, of our scientific societies. It cannot be expected that the government, even if otherwise disposed, will pay much deference to the opinion of any scientific body, composed (for the most part) of gentlemen possessing no other qualifications than general respectability, with the power of paying fifty pounds for admittance. Neither can it be expected that institutions so constituted, should employ their influence with the government, in " staying its destroying arm, in calling into action its powers of doing good, or in demanding its bounty for such distinguished men, who were especially placed under their patronage." Unless, therefore, government shall grant salaries to a certain number of its most distinguished men of science, as is done in every other country, no effectual improvement can take place. The " voluntary system," suits the state as little as the church. In return for this bounty, the society would be, as it were, the scientific advisers of the crown, they would superintend public experiments, report upon all scientific measures submitted to government, and, in short, perform

---

* *Sixty-three* of the ordinary members of this noble institution receive each an annual pension from government of 1500 francs, and the two secretaries 600 francs each.

those multifarious and valuable duties which are so admirably discharged by the Academy of Sciences of Paris. Did the nation possess a scientific tribunal of this sort, no administration would venture to act in opposition to its unanimous voice; because, in case of failure (such for instance as the loss of millions of pounds in the blundering calculations about annuities), they would be left without the shadow of an excuse for the evils that might result from their wilful rejection of the best advice.

(272.) It may be urged, indeed, that under the present state of things, our philosophers will cheerfully give to the nation the benefit of their experience and advice gratuitously, whenever our rulers will condescend to ask them; and that there exists no necessity, therefore, for burthening the national funds with the expenses of pecuniary remuneration. We believe that, to a certain extent, such patriotic feelings among our scientific men are very general, and that their desire to advance the public good will, upon most occasions, far outweigh the more sordid motives of pecuniary profit. But in the depressed state of science among us, and its neglect among the aristocracy, those who are attached to its pursuits find it necessary to follow some profession by which they can live; and, in a country like this, where the necessaries of life are so dear, and its elegancies so highly prized, abstract science, which is quite profitless, can only be prosecuted at those hours of leisure, allowed by the intervals of commercial or professional avocations. These must be followed, day after day, with undeviating regularity. How then, can it be expected, or how is it possible, for such men to withdraw themselves from the duties of their calling, for the purpose of investigating complicated questions of science, and after studying and experimenting, report upon scientific measures affecting the public welfare? They may have a strong inclination so to do; but they want the time; and time, in a commercial country, is often among the most valuable of possessions. Its value should, therefore, be paid for, if it subjects its possessor to loss; just as we should deem it unjust not to pay for any other marketable commodity. If abstract science would procure meat and drink to its possessor, if it would open a path to the esteem or the patronage of ministers, or, finally, if it was prosecuted by those who already have wealth and leisure, the case would be different; but, under present circumstances, we do not see in what manner the nation can procure scientific advisers unless by paying for them.

(273.) The next improvement, which can only originate in the government, is of far less moment than those just dwelt upon; yet it is not unworthy of our present attention, inasmuch as it relates almost exclusively to natural history. We allude to the removal of those taxes upon zoological publications now in force, and the substitution of such measures as would encourage their publication. In ordinary cases our copyright laws are not only unexceptionable, but liberal. They secure to an author the sole right of publishing his works for twenty-eight years certain, upon the presentation of eleven copies to the public libraries of the kingdom. It has been observed that this tax, " which is scarcely

entitled to that name, is a mere trifle, amounting only to the price of the paper of eleven copies, if the work is a successful one; but if the work does not sell, the tax becomes nothing, for the eleven copies have no value; and it is better for the author that they should be deposited in the public libraries than converted into waste paper. Hence it follows, that the author of a work has his property secured to him by statute, without paying for the privilege." *
Against reasons so forcibly urged, and apparently so conclusive, little, it may be thought, can be said; and yet, if applied to the case of illustrative works on natural history, that is to say, with coloured plates, they are perfectly erroneous. This error originates from overlooking the simple fact, that ordinary printed books are perfected by *one* manufacturing process, while such as have coloured plates undergo *two*, the last of these, which constitutes the *finishing*, being generally much more expensive than the first process. To bring this fact at once home to the conviction of the reader, he has only to remember, that after the expense is incurred of the setting up, in printers' types, the pages of a book, there is little more expense in printing off two hundred and fifty copies, than in printing a single one; the only additional cost is the paper, which, upon *each* copy, is a mere trifle. The book is then finished by one process, and comes perfect from the hands of the printer. But, if it is to be illustrated with coloured plates, as these plates occupy one third, or (what is very

---

* Quarterly Review.

general) one half its bulk, there is still another process to be gone into, much more expensive, at all times, than the first, and generally doubling that amount. All these plates must be coloured by skilful hands, not indeed in a wholesale way (for such an expense would be an insuperable obstacle to bringing out such publications), but as they are wanted for immediate sale. The expense of the work, since it quitted the hands of the printer, is thus increased threefold, and the author of such a work is consequently subject to a tax three times greater than if his book merely consisted of letter-press. Did the generality of illustrative works give any profit to their publishers, or even remunerate them for their first expenses, there would not be so much reason for complaint; but those who have unfortunately made the experiment, with the hope of benefiting science, if they are not absolute losers thereby, know by experience that in nine instances out of ten such publications, however admirable may be their execution, are sure to entail pecuniary loss upon their projectors. A recent case, strongly illustrating the present argument, has come to our personal knowledge, of a zoological publication, where the copies which might be claimed by public libraries were *eleven*, and the subscribers to the work were *twelve*, so that the author was subject to the expense of colouring eleven copies over and above the twelve which were sold,—an expense, be it remembered, which there would have been no necessity of incurring, except from the enforcement of the copyright laws. The sale of only *twelve* copies of each number would have done something to diminish the loss upon the publication; but when, out of these scanty proceeds, eleven other copies are to be carefully coloured and then given away, the proceeds to set against the first cost would be *nil.* It might be reasonably supposed, by any one ignorant of the actual state of science in this country, that a work which would thus fall, as it were, still-born from the press, was either utterly worthless, or at least possessed no claim to scientific excellence or beauty of execution. But such is by no means the case. The author is of established reputation, the scientific merit of the work has called forth encomiums in British and foreign periodicals, and the beauty of the plates excites general admiration. We could give other instances of similar works published in England sharing the same fate, although, perhaps, not so deplorable. Enough, however, has now been stated, to show, that, however excellent and lenient the tax imposed by the copyright act may be, upon the generality of authors; still, that on the description of works we are now considering, it is peculiarly oppressive: inasmuch as it falls with a threefold weight upon those authors who are least able to meet the demand, and who receive none of the benefits these laws extend to others. It is only when a work is eminently successful, and has an extensive sale, that piracy need be apprehended, and the enactments of the legislature become really useful. As for the value of copyright to the authors of illustrated works on natural history, it is absolutely nothing; for no one, with the least knowledge or experience in these matters, would be so inconceivably silly as to infringe the laws for the sake of incurring pecuniary loss, the certain result of such an undertaking.

(274.) The importance of the class of publications we are now speaking of, not only to the advancement, but to the right understanding of science, cannot be questioned. Words, however many, or however well selected, cannot picture to the eye the forms of things. And, next to the examination of the real object, an advantage seldom to be obtained, its correct representation is the most to be desired. Without the aid of accurate figures, natural history, in all its branches, would be involved in doubt and complexity, from the poverty of language to express the innumerable forms, and modifications of those forms, in the objects upon which it treats. So much more easy is it to impress a definite image upon the mind through the medium of the eye, than the ear, that a rough outline, a small woodcut occupying but a square inch, will accomplish this object better than a whole page of the most elaborate description. In proportion to the complication of the object we wish to make known, so is the necessity increased for calling in the aid of the graphic art. It is, therefore, absolutely essential that such works should abound in every department of zoology, because the objects to be made known by such means pour in upon us from all parts of the world, while the difficulty of discriminating them, by mere words, is proportionably increased. But by whom are such works (necessarily expensive from the cost of the labour to which they owe their excellence) to be encouraged or patronised? The natural supposition would be, by those institutions or

IMPORTANCE OF ILLUSTRATIVE WORKS.

societies expressly formed for the advancement and protection of science, — by the nobility and the wealthy of the land, who feel a laudable pride in the richness and excellence of their libraries, — and by men of science, who are themselves interested in the success of their favourite pursuits. But, if these sources of patronage are ineffectual, it is clearly incumbent upon a liberal government, zealous for increasing the facilities of knowledge, to stretch forth a protecting hand, and either directly to take upon themselves the cost (with proper limitations) of publishing such works as, upon mature consideration, may be deemed worthy of national aid, or indirectly, by other means, give to those authors who will take the risk upon themselves certain privileges or immunities, proportionate to the hazards they incur. All this will doubtless appear preposterous to those who think that science is to be advanced by the cheap compilations of the penny press, which, we feel almost ashamed to say, are now the only books upon natural history which suffice to please the "great taste" so much talked of as existing in the public at large : but the opinions of such persons can never controvert the well-known fact, that the governments of other nations make especial provisions for assisting in the publication of expensive works, which, without such aid, would never see the light. France, who seems determined to take the pre-eminence in all questions of national science, annually appropriates no less a sum than *ten thousand pounds* to the costs of publication and subscription to scientific works, nearly the whole of which relate to different departments of natural history. By this munificence she has been enabled to give to the world all the zoological discoveries made on her scientific expeditions, besides aiding the exertions of her naturalists at home, by enabling them to bring out a series of illustrated folio volumes, no less valuable than splendid, at once attesting the liberality of the government and the judgment it has exercised. How much can be done, even with a tenth of the above sum, by judgment and discretion, has been made apparent by the publication of the *Fauna Americana Borealia*, containing the zoological discoveries made in Franklin and Richardson's expeditions; and which, but for the grant of 1,000*l*. towards the engraving of the plates, would most assuredly never have seen the light. Were the government fully aware of the extent of the satisfaction felt by all the well-informed classes, at this *maiden* act of generosity, small though it was, that even in the periodicals and newspapers of the day there was but one opinion expressed, at the very time when the cry for economy and cheap government was at its height, — if the present administration, we repeat, were but fully aware of all this, they would never again hesitate to propose similar or even more liberal grants, and thus gather to themselves those "golden opinions from all sorts of men," which the country has long wished to give, and which its science has so long languished for.

(275.) But the occasions for exercising such liberality on undertakings connected with national discoveries, are few and far between; and we cannot hope, in this our generation, to witness direct patronage extended to private undertakings, however excellent or however important. Yet there are some measures, of an indirect nature, which would materially promote the same object, and soften the peculiar disadvantages attending the authors of costly publications. In the first place, those who wish it might be exempt from all the advantages and the penalties of the copyright act, and the government, instead of claiming eleven presentation copies, might, without a charge of great extravagance, subscribe for an equal number, and present them, as a gift of the crown, to the chief public libraries of the nation. In the next place, let a drawback be allowed on the excise duties paid on the paper consumed, provided it amounts to a certain given sum, and is of such a description as to show, at once, that it has been used for a large sized and expensive work. These two concessions, simple and practicable as they undoubtedly are, would at once have a powerful effect on the publications in question; and this in two ways: first, by diminishing the original cost, and consequently removing the great obstacle to their extended sale, now existing in their high price; and secondly, by thus enabling the publishers to find a market for works of this description on the Continent, where the high price of those published in Britain acts almost like a prohibition to their sale. It is a well-known fact, that the costs of publishing a book in England are *exactly double* what they are in Paris; a difference easily explained by the heavy duties upon our paper, and the higher wages of English printers. It therefore follows, that the high price of English books, but more especially those of which we are now speaking, almost excludes them from the continental market, while nobody will buy them at home. The only plan, therefore, by which this virtual prohibition could be overcome, would be by diminishing the cost of production in some such manner as we have already suggested, or by allowing a small bounty on their exportation. The fact of the matter, however, is this; that for books of this description there is, in all countries, more or less, such a very limited demand, that profit is entirely out of the question, and the only effectual way of promoting their sale, and of reducing their present cost, would be by a general agreement, among all civilised governments, to admit them free of all import duties. We believe some such liberal measure has been adopted in France, and we trust that the American, if not our own government, will not be tardy in performing this small act of generosity to men of science, who are generally compelled to publish at their own cost and charges, from the universal disinclination of the commercial booksellers to embark their capital in such hazardous projects. The import duties in America are so heavy, that illustrative works, printed in England, can find no purchasers among our Transatlantic brethren, distinguished, as they undoubtedly are, by a much more national encouragement of science than exists with us, and where natural history, even in some of the most remote provinces of the Union, already has its regular professors. Petty jealousies, in such matters, if they really exist, ought surely to be laid aside; where no profit worth naming can be derived, the idea of competition is perfectly ridiculous. We should hail with pleasure every addition made to our stock of knowledge by the press of America, while, by admitting their publications free of duty, we should doubtless receive the same indulgence: and the costly works of British naturalists, with a small reduction of their original price, and disburthened of duties on their arrival, might then find purchasers in large districts of America, where at present they are only known by name.

(276.) 5. The last subject connected with our present enquiry possesses much interest in itself, and still more from its recent discussion in the House of Commons. It is on the propriety or impropriety of conferring honorary titles or distinctions upon those of our philosophers who have benefited their country by their discoveries or inventions.

(277.) That distinguished merit, of whatsoever description, should receive reward, either pecuniary or honorary, proportionate to its nature and degree, no one will deny; and that great intellectual acquirements are far superior to qualities derived from the exercise of animal faculties, is also an undeniable truth; proved, if proof were necessary, by the rarity of the one and the frequency of the other. But the bulk of mankind are but little influenced by abstract truth. They will assent to its doctrines from their incapacity of denying them; but they will seldom carry this assent into practice. They will be content to admit the general principle; but, if it is to be applied to particular cases, they shelter themselves under the common excuse, that custom, or fashion, or national feeling, is against the measure; they look for precedents; they cannot go against public opinion; they must have the sanction of example, before they can make innovations. Before, therefore, we analyse the question above stated, it may be as well to scan the sentiments of nearly all civilised nations, proved by their acts on this subject, that we may at once see whether they accord with those which have hitherto guided the government of this country in the honorary reward of merit.

(278.) A few striking facts, in addition to those noticed in the preceding chapters, are alone necessary to establish this proposition, — That from the earliest revival of science down to the present moment, the governments of the most powerful and enlightened kingdoms in Europe have considered honorary titles or distinctions appropriate rewards to men of science; and that in several we find distinct orders of knighthood, or of merit, expressly instituted for this purpose.

(279.) We must begin with France, because, as her institutions are better known in this country, her situation offers an immediate point of comparison, while the industry of a recent author enables us to argue from the most convincing of all proofs — names and figures. Mr. Babbage has given the following tables, the correctness of which has not been questioned. " If we analyse the list of the Institute," observes our author, " we shall find few who do not possess titles or decorations; but, as the value of such marks of royal favour must depend in a great measure on their frequency, I shall mention several particulars, which are probably not familiar to the English reader.

	No. of Members of the Institute of France who belong to the Legion of Honour.	Total No. of each Class of the Legion of Honour.
Grand Croix	3	80
Grand officier	3	160
Commandeur	4	400
Officier	17	2000
Chevalier	40	not limited.
	67	

Here then we have no less than *sixty-seven* members out of *seventy-five* upon whom the government have thought fit to bestow titles of distinction. Our Royal Society, which corresponds only to the French Institute in being the best of its kind in Great Britain, contains 685 members, among whom there are only two or three Hanoverian knights, who have very recently been so created, in consequence of their scientific attainments.* Next, as to the nobility belonging to the Institute, the greater number of whom, be it remembered, are men of science, have contributed to its annals, and have been rewarded with these titles in consequence.

Dukes, 2.   Counts, 4.   Barons, 14.
Marquis, 1.   Viscounts, 2.

making a total of twenty-three, five of whom are peers of France. Our list of the Royal Society can probably show an equal, if not a superior number of

* Titles which are the reward of members eminent only in the medical profession do not, of course, bear upon the present question.

titled names; but it is scarcely necessary to add, that not one has been the reward of intellectual acquirements, while of those in our list who have contributed to the Philosophical Transactions, and have thereby demonstrated their scientific knowledge, there appears to be only *one*."—" It must not, indeed, be inferred," observes our author, " that the titles of nobility in the French list were *all* of them the rewards of scientific eminence; yet many are known to be such; but it will be quite sufficient for the argument to mention the names of Lagrange, La Place, Berthollet, Chaptal," and, last, though not least, Cuvier.

(280.) Need we, after such facts as these, search for further details on the decorations and orders of merit bestowed upon living philosophers by Prussia and Bavaria, which, at the present moment, of all the European nations, are not inferior to France, in their munificent encouragement of science, — by Saxony, whose chief astronomer is likewise an ambassador, — by the Grand Duke of Tuscany, whose prime minister is a celebrated mathematician, — or by Russia, whose Aulic councillors are almost exclusively chosen from the ranks of science? These instances, taken at random, are sufficient indications of the state of feeling throughout Europe on the question before us, — that philosophers should have the option of accepting those titular distinctions, which are so much coveted by the bulk of mankind, and so profusely lavished upon others. Their possession, it is true, by such men, can add little or nothing to that imperishable fame which is their chief desire; a paramount feeling which prompts them to the sacrifice of personal interest, of wealth, and of domestic comfort.

(281.) If, therefore, a question is to be decided by the general custom and feelings of all civilised nations, we might here stop, and enquire, what reasons can be assigned by Englishmen for thus pertinaciously refusing to follow the example of the rest of the world? Why are we to refuse specific honours to *one* class of public benefactors, when we bestow them, with lavish profusion, upon all others? Is this eccentric opinion held by the nation at large, or is it peculiar only to the executive government? Let us see how this case stands. There is a sterling good sense and love of independence in the people of England, which leads them to rejoice in the success of any one, who, by sheer force of personal merit, gains distinction or reward. Now this feeling springs from two motives, which constitute the most prominent ingredients in the national character: the one is a love of justice, the other of independence; the first fostered by the excellency of our judicial laws, the latter by the freedom of our constitution: with these is blended a third, which tells us, that if national honours attended great excellence, distinction may be attained by those who, possessing talents, exert them to the utmost. An instance, therefore, of independent and self-created merit, fitly rewarded, comes home to the individual feelings of every good man, and leads us to extol it as an act of justice. Nothing illustrated this proposition more strongly than the praise which was bestowed, almost to extravagance, on the recent elevation of two or three of our philosophers to the honour of knighthood, and the bestowal of a small pension upon another.* At a time when the daily press teemed with invectives against the titled aristocracy, the existence of sinecures, and the granting of pensions, all parties in the kingdom, from the philosophic congress held at Cambridge down to the most violent of the radical papers, united with one heart and one voice in extolling these acts of national liberality, and in lamenting they were so seldom exercised. It is plain, therefore, that the nation at large, so far from participating in the indifference habitually shown by our government to the science of the country, applauds and approves of those honours occasionally bestowed upon its votaries. From whom, then, does this injustice proceed? Certainly not from the nation at large, or from its intellectual classes. Mr. Babbage has made it a subject of bitter complaint, and the Quarterly Review has, to use its own emphatic words, "unfolded a series of grievances of the most afflicting kind." Murmurs and reproaches have spread wider and louder in proportion as they have been disregarded, until we now find them bursting forth in the parliament of the nation.

(282.) On the recent discussion in the House of Commons as to restricting the Order of the Bath, a well-known member, decidedly opposed to all unnecessary expense or unmerited distinctions, is reported to have made the following sound and admirable remarks:—" Although England was strictly a

---

* See Proceedings of the British Association at Cambridge.

DISCUSSION OF THE SUBJECT IN PARLIAMENT. 413

civil country, — a nation opposed in constitution, habits, feelings, and desires, to the despotic states of the Continent, still it was found that all distinctions were almost exclusively conferred upon military and naval officers. There were no orders to reward the man of genius and of science; none to pay him for the halo he gave to the name of England. There were few men of eminence (whose pursuits were civil) Knights of the Bath; they were, indeed, few and far between. Although there were five orders, — the Garter, the Thistle, St. Patrick, the Bath, and the fifth, a new one, St. Michael and St. George of the Ionian Islands, yet there was not one to give honour to the man of genius and talent. It was quite different on the Continent; *there* was found the man of literary eminence in his proper place; if they went to Prussia they found him honoured; if to France, highly and nobly distinguished. *England stood alone, and yet she was a civil country.** (*Hear, hear*)."

(283.) This speech, which in our opinion places the inconsistency of our honorary distinctions in a new and forcible light, as being utterly at variance with the nature of our government, was answered by another distinguished member, who in the late administration was one of the chief advisers of the crown. This reply deserves great attention, not only from its nature, but as revealing to us those sentiments entertained in the highest circles, which have hitherto guided the government on all questions influencing the science of the country, and from which has resulted its present humiliation. The

---

* Taken from "The Sun," April 19. 1834.

right honourable member in question, in reply to the foregoing speech, is reported to have said, "that there were some men to whom honours were unnecessary, and which could not confer higher dignity than their genius and their talent had invested them with. What, he would ask, could a blue riband or a collar do for a Newton? Would they make his name more hallowed, — his family more endurable? No, certainly not. There should be a line of distinction drawn, for, if not, many would be seeking them. He would not mix up scientific ingenuity with military favour; he would leave it to the possession of its own ennobling honours."

(284.) Had the honourable member who spoke next in the debate, — himself a man of science, and a vice-president of the Royal Society, — whose sentiments are known to be in unison with those expressed in these pages — had *he* replied, and combated what may well be deemed the unsoundness of these doctrines, he could at once have exposed their fallacy, and have spared us the ungracious task of animadverting upon one whose talents we admire, and whose eloquence can give to any cause he espouses almost the force of demonstrative conviction. It is seldom that an opening occurs, in parliamentary debate, for the discussion of matters affecting the interests of science; and when so few of our representatives are qualified to speak on such things, it is doubly disheartening to witness such opportunities for vindicating her rights neglected. Perhaps, however, the forms of debate might have prevented our vice-president from following up the subject; a duty, however, which is imperatively imposed upon any one who complains of the " decline of science."

(285.) The speech last quoted contains three assigned reasons for showing the impropriety of conferring titles of honour upon scientific men: — 1. Because their highest dignity being the possession of genius and talent, *therefore*, national honours were unnecessary; 2. Because, if such honours were conferred " many would be seeking them;" and 3. Because scientific merit " should be its own reward."

(286.) There are some questions so long set at rest by the general voice of mankind, that to enter elaborately into their defence, on ordinary occasions, is not only superfluous, but may become nearly as ridiculous as to fight with our own shadow. Yet if these very opinions, long explored by the reflecting, are taken up for some particular purpose, clothed with eloquence, and delivered with grace, by an accomplished orator, they are listened and assented to by the assembly, who applaud that which, if spoken by a common person, would immediately produce ridicule. Of this character is the opinion that where great merit exists, no outward sign or symbol — by which its possession is to be made known to the world — is at all necessary. Or, if necessary in *one* class of excellence, it is not so in another. An ordinary person who would thus argue, must either be very little acquainted with human nature — with the general sense of mankind on this subject — or he must imagine that the feelings of philosophers are totally different from those of other men. He must suppose that men of science are wholly exempt from the influence of one of the ruling passions of our nature—ambition. That such men can derive no personal gratification from such marks and tokens as at once show to the world the gratitude of their country, the esteem of its ministers, or the favour of its sovereign; that they are quite indifferent to the public acknowledgment of their merit; that they are well content to be received in society on the same footing as the most illiterate citizen; and that, in the "pride of place," they have not the least sense of humiliation in being jostled in the crowd, and ordered to make way for a city alderman, carrying up an address for which he is to be knighted. Now all these are the inevitable consequences of withholding national honours from men of science. What are such honours made for, but to be given to those who deserve them? Why are they created, but that the nation should know to whom it is indebted for its glory? If such honours cannot confer higher dignity, in a *worldly* sense, than genius or talent, of what possible use do they become? they are altogether as worthless to the warrior, the statesman, or the nobleman by birth, as to the philosopher. What, it has been asked, " could a blue riband or a collar do for a Newton? would they make his name more hallowed, his family more durable?"—What, let us in turn demand, can a multitude of ribands, and crosses, and collars, do for a Wellington; will they make *his* name more famous, *his* family more endurable? The answer to both has been already given, " No, certainly not." What, then, is the use of such things?—baubles though they be. The answer is obvious—they evince the gratitude of a nation for benefits conferred. They certify to the world that their possessors have reached the height of their ambition; an ambition which is felt by the sage no less than by the hero, which is the noblest stimulus to exertion — the most hallowed feeling of the human heart; they are at once enrolled and distinguished as *benefactors to their country*, if not to the human race. Such is the legitimate use of titular honours and their accompanying signs. That they are perpetually degraded to ignoble purposes, to reward the courtly sycophant, the political apostate, or the wealthy imbecile, is most true; but were they *invariably* so perverted, it is obvious that no good man would accept or desire them. They still retain the lineaments of their original image, and their superscription can still be traced; and who can imagine that philosophers only are to be insensible to such things, that *they* only are devoid of ambition, that *they* only are indifferent to the applause of their countrymen, few, very few of whom would admit their deserts, and still fewer understand them, but for such public testimonials? To argue, therefore, against the existence of such feelings is to suppose there is no ambition in the world beyond that which belongs to the grosser and baser passions of our nature, which strives after wealth, or power, or possessions, for themselves alone, totally regardless of those means they supply of doing good to others. Philosophers are but men, generally exempt, indeed, from the vanity and pride of vulgar minds; yet still they cannot be insensible to distinctions, earned by intellectual exertion; any more than the warrior or the statesman can be supposed to despise the decorations of a riband, or the homage of a political party.

(287.) But then it is urged, if philosophers were to be rewarded *by titular distinctions,* " many would be seeking them." No doubt, many aspirants would be found for these, as there are for all other honours, whose qualifications were trifling. But in what manner this evil is to be peculiar to the class of society under consideration has not been explained. As national honours, of whatsoever description they may be, are highly and deservedly prized; so, as a matter of course, will they be sought for and coveted. When a vacancy occurs for the decoration of a blue riband, are there not many who seek to fill it? Those who are in power can best answer this question; but every one acquainted with human nature, knows by induction that there must be a host of aspirants, where honours are to be gained; while to suppose such rewards were not of sufficient value to excite competition, and induce " many to seek them," is at once to proclaim their unfitness for *exciting that emulation which is one of their legitimate uses.* But supposing, for a moment, that England, like all other civilised European nations, had her own order of merit, by whatsoever title it were called, and solely restricted to her philosophers. Can it be imagined, for a moment, that there would be a tenth part as many aspirants to its honours, as there are now to the titles belonging to the existing orders? Unquestionably not, and for this simple reason: hundreds, we might almost say thousands, in a country like this, from the possession of wealth, rank, or

DIFFICULTY OF FILLING A SCIENTIFIC ORDER.

connection, joined to the qualifications of personal courage or legal knowledge, are at this moment as much entitled to be made knights of the established orders of merit, as most of those individuals who have actually been so honoured. No one will deny this, because it has tacitly been admitted in parliament. With science, however, the case is totally different. Instead of such a superabundance of qualified individuals for filling up a scientific order, the real fact would turn out to be, that if high excellency was *alone* regarded, government would find great difficulty in filling up the ranks. Instead of being embarrassed where to decide in the multiplicity of equal claims, they would be perplexed in finding men sufficiently well qualified; and if they limited the number of the order even to *fifty*, they must of necessity admit many who now occupy only a second or a third station in the ranks of philosophy. This objection is, therefore, a peculiarly unhappy one, since the danger to be feared is, not the difficulty of *selecting*, but the difficulty of *finding*. Every one, at all acquainted with the subject, and with that description of excellency which is possessed by titled philosophers upon the Continent, is fully aware of the paucity of such scientific attainments in Britain. And even common observation will show that the numbers among us, who pursue the higher walks of science, " are very few, and probably will long continue so."

(288.) We fully agree, indeed, with the right honourable member from whom these objections have originated, in the justness of his remark that "some distinction should be made*," and that those outward signs or titles, which might be the reward of our scientific benefactors, should not be too common. Let them, therefore, be of such a description that they will be inaccessible to the army, the navy, the bar, the church, and the medical profession, unless individuals belonging to these professions, also possess scientific attainments. Under such restrictions, there is no fear of scientific honours being too common, if those only who deserve them are so distinguished. But once admit *all* these various professions, and the honour becomes nominal, and it will no longer be an object of ambition to the man of real knowledge. That some discrimination of this sort should be made, is abundantly evident. The truth is, that these professions already mono-

---

* The government would have done well, perhaps, if they had followed up this principle of preserving "distinctions" in the lavish profusion of honours stated to have been conferred of late years. "Since the peace of 1816, no fewer than 97 Knights Grand Crosses, 164 Knights Commanders, and a whole regiment of Companions of the Order of the Bath, have appointed: these are all military and naval men; and though the order does admit the civil servants and benefactors of the state, yet only 15 of this class have been appointed, and *not one* of these knights are men of either science or literature. In the long list of knight bachelors, we meet with a singular assemblage of characters. Judges, lawyers, soldiers, sailors, physicians, surgeons, apothecaries, painters, architects, booksellers, and quack doctors, and all the *operatives* of the political machine are marshalled in ludicrous juxta-position. A few honoured names, indeed, grace the multifarious list," but not more than two scientific characters are to be found. Quarterly Review, p. 332.

polise *all the honorary distinctions of the empire.*
They possess titular distinctions peculiar to themselves, which at once attest and define the rank of each individual occupied in his own profession; but, as if they only deserve honours, they are selected to recruit the ranks of our nobility, and to share among themselves all those tokens of outward dignity, which the laws and customs of other nations throw open to excellence of every sort. The professional titles belonging to the army and the navy fix at once their station in society, and these are the proper and legitimate rewards they should have. So also in the bar and in the church, there are titles and dignities of the same description, differing, indeed, as they rightly should, in name, but equally marking the different gradations of estimation or of preferment, to which those who enjoy them have respectively attained. With these appropriate distinctions let them be satisfied; or, if one or two orders of knighthood are instituted for conferring additional dignity upon the possessors of animal courage (we use not the term reproachfully), let there at least be others, equally set apart for those who have achieved the most glorious of all victories — the victory of knowledge over prejudice; whose conquests have at length seated science and civilisation upon the throne of Europe, formerly occupied by barbarism and ignorance. *This* is the distinction which should be drawn; a distinction as great as that between matter and spirit, between the arts of war and the arts of peace. We deny not that both these qualifications are essential, in the present condition of the world, to the prosperity of a state,

but we also contend that both should be equally honoured and rewarded by the nation.

(289.) Let us now look to the consequences of leaving scientific excellence in the " possession of its own ennobling honours." And why not leave all other excellence to the same fate? Does science with us, as with other nations, feed her children with the necessaries of life? Does she make her ways the ways to estimation or preferment, to favour or to patronage? If she did, her votaries might then be very well " left to the possession of their own ennobling honours;" they would have no cause for complaint, they would then enjoy the substantials of greatness, and would care very little for its nominal privileges. But what is notoriously the result of this system — this visionary scheme — by which science is so respectfully neglected? It is, in Britain, to come into an heritage of poverty, obscurity, and neglect. To use the words of an eloquent writer, " He whom the Almighty has chosen to make known the laws and mysteries of his works — he who has devoted his life, and sacrificed his health, and the interests of his family, in the most profound and ennobling pursuits — is allowed to live in poverty and obscurity, and to sink into the grave without one mark of the affection and gratitude of his country."* Such is the " barren heritage" which the ministers of this country would assign to her philosophers. Such are the " ennobling honours," of which they are to be left in possession.

(290.) And why is this? why are excuses sought

---

* Quarterly Review.

for thus acting in opposition to the plainest dictates of reason and of justice, to the feelings of all other civilised governments, and to those even of the intelligent classes of this country? It is not because the value of science is unknown to our rulers; for they have, as it seems, so enlarged a conception of its worth, that they imagine it can provide for itself. It is not because the nation at large disregards it, because in no other country are there so many public associations for its culture. No. It is because science enters not into the strife of politics, she " brings up no reserve to the minister, to swell his triumph or to break his fall;" she remains passive in the warfare of elections, she possesses no courtly influence, and flatters no courtier; she neither comes recommended by wealth, by power, by titles, or by interest. She has, in short, nothing to give; *and therefore it is, that she has nothing to receive.* All other reasons for her neglect merge into this; and it may be fairly questioned, whether her votaries, " now depressed to the level of hewers of wood and drawers of water," will ever recover their caste, until those who direct public affairs will abandon the line of argument we have here attempted to refute. These arguments have been investigated more closely, perhaps, than may appear necessary; but as emanating from a late cabinet minister, they deserved every attention, since they clearly show, in language not to be mistaken, the sentiments which pervade that particular school of politics which has hitherto been paramount in the national councils.

(291.) And yet, so far from desiring to see a new order of merit at once instituted for our philosophers, it may be questioned whether such a measure, in the *present* state of affairs, would be in any way expedient; for, if we are retrograding in the higher walks of science, the *first* duty of a wise administration will be to check that declension, and provide for a restoration. Let us first encourage and foster scientific talent, and then, if it bring forth good fruit, let it be honoured and rewarded. At present our scale of excellency is at so low a standard, that with the exception of about a dozen names,—and these confined to two or three particular departments,—we have really so few possessing those high qualifications which are rewarded by honorary distinctions in other countries, that if a distinct scientific order were instituted, there would be a lack of members to fill it! The inevitable consequence of this deficiency of real merit, would be the admission of many of very slender pretensions; while—from the incompetency of the higher classes to judge for themselves on such matters, and the propensity there is in all our administrators to augment their political power—the new order would be chiefly filled through the channels of patronage, and by amateurs rather than by acknowledged adepts. A reserved and retired disposition, absorbed in its own unobtrusive pursuits, and shunning those busy haunts where personal popularity is to be gained, and personal interests advanced, are the general characteristics of the man of real knowledge. These men are the very last who are likely to gain the attention, much less the regard, of those by whose recommendation such favours are dispensed. Should any administration, therefore, in future times, really wish to place the science of

Britain upon another footing, it must begin with national institutions for instructing, and pecuniary endowments for maturing science; and, lastly, it might then proceed to reward those who attain pre-eminence.

(292.) We thus bring our argument to the point from whence it began; namely, that a love for natural science must be imbibed at our universities; where, to be taught effectually, it should be incorporated as a necessary part of academic education. It is obvious, that tastes so acquired will have a powerful influence on the minds of those who may hereafter become legislators; and that finally, a government composed of such legislators, will feel a personal interest, far stronger than that of political, in seeking out and rewarding, both with the pensions and honours of the state, those whose names are the brightest jewels in the diadem of the British empire. It will be only when the founts of science are opened to the sons of our nobility and aristocracy, at those venerable and noble institutions where they are educated; it is only *then* that we can expect to see the philosophic spirit of the Boyles, the Cavendishes, the Montagues, the Willoughbys, and the Howards of former and better times, again revive in their descendants, and once more occupy their proud station in the scientific annals of England. It will be only then, that the honours of the state will be thrown open to our philosophers and literary characters. Then will the sage and the hero, as in other kingdoms, deliberate in the same cabinet; they will be associated among the privy councillors of the king, sit together in the united parliament, bear the same titles, be decorated with the same orders, and the mind and the arm of the nation will be indissolubly united for its glory or for its defence.

(293.) We cannot altogether abandon the hope, that at a period unexampled in our history for the diffusion of knowledge among the people, in a time when the name of Brougham will be inseparably connected with this new era of intellectual developement, and that not as a private individual, but as the Lord High Chancellor of these realms, possessing rank, power, learning, and eloquence, all that is necessary, in short, for conceiving and executing the most noble designs — we cannot abandon the hope, that something effectual may yet be done, even in these our times, to remove the stigma that has so long rested upon our national character. We might suggest to that exalted individual a truth which he will at once perceive, that unless the spring-heads of knowledge are sedulously repaired and renovated, the stream will be soon exhausted; and that in proportion as we may anticipate a demand for more and more information, we cannot furnish that supply unless we sedulously protect those few secluded founts whence alone it will gush forth. While we are indefatigable in diffusing that knowledge which is already possessed, let us be equally careful in creating a fresh supply, to be poured forth abroad when that which we have in keeping is exhausted. Without such prudence, it is not difficult to foresee the injurious effects which will follow; for the science of the country already begins to show them in its declension. Knowledge, indeed, will be diffused, but it will become proportionably superficial: all that is light, and novel, and amusing, will be eagerly caught hold of, and scientific trifles will take the place of scientific inductions.

(294.) But if this our effort fail to rouse the attention of the present administration, " we must wait for the revival of better feelings, and deplore our national misfortune in the language of the wise man : ' I returned, and saw under the sun that there is neither yet bread to the wise, nor yet riches to men of understanding, nor yet favour to men of skill.' " *

---

* Quarterly Review, p. 342.

CHAPTER IV.

SUGGESTIONS FOR THE REFORM AND IMPROVEMENT OF OUR SCIENTIFIC SOCIETIES.

(295.) WE have enlarged, in the preceding chapter, upon those means possessed by the government and the universities for giving a new impulse to the science of Britain; because no renovation can be complete and effective, which does not commence from these sources. The organisation of our scientific societies, however, is a subject of some moment; because, unless we ourselves evince a disposition for improvement, we cannot expect assistance from higher quarters. To expose defects, and to animadvert upon the proceedings of such institutions, is at all times an ungracious task; yet experience has shown, that it is most necessary. Imperfection attaches to every thing human; and we are most ignorant of that imperfection, in proportion as we turn away from advice, and disregard the opinions of others. Now, where there is a disposition in the majority to think correctly and to act wisely, we are more disposed to treat with indulgence existing defects, than to expatiate upon their universal prevalence; preferring, at all times, dispassionate reason to bitter declamation and general sarcasm. We, therefore, leave to others the exposition of existing or assumed abuses; contenting ourselves with touching only upon those points which regard the well-being of our respective societies, which are most conducive to effect the objects they have in view, and which are sanctioned both by reason and experience. With these feelings, we shall now proceed to offer a few remarks on the chief metropolitan societies and institutions formed for the promotion of natural sciences, and more especially natural history: viz. the Royal Society of Great Britain, the Linnæan Society, the Zoological Society, and the Entomological Society. The Geological Society will be altogether omitted: first, because it more concerns the mineral than the animal kingdom; and, secondly, because its laws and its management appear to be so admirable, that they may be looked upon as a model for all others.

(296.) The defects in the management of the Royal Society have already been touched upon; they have been treated of in more detail by Professor Babbage, and have been intimated by Sir James South. There is one censure, however, brought against the society by the Quarterly Review, which may be here repeated, as it is passed not merely upon the Royal, but upon all the leading societies of London. It is, that " they have not employed their influence, with the government, either in staying its destroying arm, or calling into action its powers of doing good, or in demanding its bounty for those distinguished men who were especially placed under its patronage." * But this censure, just though it be, attaches more to

---

* Quarterly Review, p. 330.

those members of the government and of the aristocracy who are *fellows*, than upon the *councils* of these societies. The latter, being chiefly composed of untitled, and therefore uninfluential, men, would naturally feel timidity in making representations which they had not influence to support; whereas the former, being chosen more for their political influence than for their scientific attainments, were certainly bound, in duty, to use that influence which procured them their admission, for the benefit of science and of its professors.

(297.) We pass over the obvious expediency of free discussion, and all those ordinary means for insuring the honest and faithful administration of the pecuniary affairs of these societies. The chief point at which the *Royal* Society should aim, is that of rendering it an object of ambition among men of scientific eminence, to be enrolled among its members. All will admit that this is most desirable; the difficulty lies only in the means by which it can be accomplished. It appears, from the statement of Professor Babbage, that some time ago, " many of the more scientific members felt that some amendment was absolutely necessary to the respectability of the society;" and a committee, in which we find the names of Wollaston and Herschel, was accordingly formed. " The council received their report at the close of the session; and in recording it on the journals, they made an appeal to the council for the ensuing year to bestow on it *their earliest and most serious attention*." It appears, however, that for some unassigned reason, this strong recommendation was never attended to, and the matter was suffered to drop. Nothing could be more injudicious than such a step, or more disrespectful both to the eminent men who composed the committee, and to the members at large. If the measures so recommended were impracticable, we should have been told so; if otherwise, they should immediately have been acted upon. It is a great pity that the opinions of a committee, so composed, concerning the most vital point of interest to the society, namely, its scientific respectability, should have been kept secret; since it is difficult to imagine that *some* of the improvements, at least, which they recommended, might not have been carried into effect. Ignorant of that document, we must therefore enquire how far the object of respectability would be attained by the different means that have been suggested. 1. By the " ejection of useless fellows;" 2. By their restriction; 3. By their division into two classes; and 4. By especial regulations for future admissions.

(298.) The first plan of *ejection*, proposed by a council of the R. S., in August, 1674, wherein was Sir Christopher Wren, however honest and just, is quite inapplicable to the conciliatory principles of proceeding of 1834; and it may, therefore, be dismissed. The plan of *restriction* originated with Dr. Wollaston, who thought that the society should be limited to four hundred: this expedient is quite as impracticable as the last; for as the society now consists of more than seven hundred, there must be a complete suspension of fresh admissions until upwards of three hundred of the present race have died. Besides, the mere limitation to any given number brings with it no accession of scientific honour, if members are to be admitted upon the same easy terms as they are now. Besides, those who would raise the reputation of the society, must benefit us by their *head* more than by their *purse.* The third is a suggestion of Mr. Babbage, and is by far the most simple and practicable plan yet promulgated. It proposes that, in the printed lists of the Royal Society, a star should be placed against the name of each fellow who has contributed two or more papers which have been printed in the Transactions, or that such a list should be printed separately at the end. The immediate effect of printing such a list, it is urged, would be the division of the society into two classes. Now, if the working class — which would of course comprise those whose names constitute the honour of the society — were to be distinguished by a separate designation (as that of fellow in opposition to member, or otherwise), the only real objection to this simple plan of proceeding would be done away with; but without a more marked distinction than an asterisk, a dagger, or those conventional signs used in printing, it may be fairly questioned, whether, if the higher class be not more plainly defined, it would become a matter of ambition to belong to it? As for the " great objection" put forth against such distinctions — that they would be displeasing to the rest of the society — I really think it too trifling for discussion, especially after what has been said upon it in another place.* Trifling, however, as such an ob-

---

* Decline of Science, p. 156.

jection is, in the mind of any one who has well considered the subject, it would, doubtless, be sufficient to prevent the execution of such an arrangement. It, therefore, only remains to consider the practicability of the fourth plan, which would, in process of time, purge the society of all "*useless fellows*," and thus restore it to its pristine vigour. To effect this, let no future members be considered eligible until they have given in an essay or paper to the society, in their own line of science, by which their fitness and proficiency can be fairly judged. Some few exceptions to this general rule might be made, as in the case of philosophers, whose works have already procured them high reputation; or of noblemen, who are acknowledged patrons of science or of learning. By these regulations the present members, worthy or unworthy, will be left in quiet possession of their " vested rights;" and no offence will be given to either. Admission to the society will immediately become an object of ambition from the very moment the new law is promulgated, and the next generation would see the society assuming that elevated station in the ranks of European science, which is her legitimate right, concentrating within herself nearly all the varied talents of the nation.

(299.) It is clear, however, that even this last regulation would be much less effectual than it should be, unless some considerable reduction was made in the amount of fees paid for admission. These amount, at present, we have been told, to no less than 50*l.*; whereas, some fifteen years ago, we remember to have paid only about 30*l.* We know not, nor is it important to know, what were the reasons for this additional tax being laid upon scientific aspirants. It may have had its advantages, but the evils it has produced have far outweighed them. It is well known how very few of our philosophic enquirers are men of such independence as to render the payment of so large a sum otherwise than inconvenient, if not impossible. While those who have already built up a reputation for themselves, unaided by scientific titles, can derive no additional honour from being a fellow of this or that society; and, therefore, even if their circumstances are easy, they never think of expending 40*l.* or 50*l.* for such an unprofitable purpose. Now, what is the consequence of this feeling? Several of the highest characters in the science of Britain do not belong to the Royal Society; and this for two reasons: first, because the payment of such high fees is inconvenient; and, secondly, because scientific excellency is not an indispensable requisite for admission. On the other hand, the aristocracy of wealth, who, in this country, measure the value of every thing by what it costs, readily pay their 50*l.*; and imagine, that to associate with philosophers is to imbibe a portion of their reputation. We firmly believe that this is one among the many causes that have operated of late years to the disadvantage of the society, whose ranks, formerly recruited from the republic of science, are now chiefly filled up by the aristocracy of wealth. That it has increased its respectability of station, according to the vulgar idea, derived from wealth, there can be no doubt; but it has certainly diminished its reputation for science, inasmuch as it virtually excludes many of the highest fame, who would otherwise, in all probability, be enrolled amongst its members, and give to the Society that honour which is the vital spark of its existence. And why should these enormous fees be continued, not merely by the Royal, but by the Antiquarian and the Horticultural Societies? Why should they be higher than those demanded by the Astronomical and Zoological, or even by the Geological or Linnæan? Each of these societies publish their Transactions as frequently, and are subject (as far as we know) to the same necessary expenses for their management; and surely, when so sore an evil has been hinted at, as it would appear, from the president's chair *, it behoves the council to take the matter into their serious consideration.

(300.) We neither know, or much desire to know, if any and what steps have been taken to remedy those defects of internal administration, charged against the Royal Society, by those whose opinions are already before the public. We are satisfied that an efficient reform must commence with removing great grievances, and then proceeding to the lesser subjects of complaint. When it is considered, that the society, as a body, has little or no political influence (for it could not save the board of longitude), and that its executive members, from deriving no pecuniary recompense from the government †, are obliged to give

---

* Anniversary address of H. R. H. the President, for 1832, p. 31., concluding sentence.

† The secretaries only have a small stipend — much too small — paid by the society.

their chief attention to their own affairs, we are well disposed to pass over with indulgence many of the petty charges not yet substantiated. There is, nevertheless, one subject that has been put forth against it, upon which, as being intimately connected with natural history, we shall venture to touch. An impression has long existed among the naturalists of this country, that their favourite science, although not professedly, had been virtually excluded from those to which, of late years, the Royal Society had more especially restricted its patronage and encouragement; and this implied understanding had arisen from the institution of the Linnæan and, more recently, of the Zoological Societies, both of which were formed more particularly for the advancement of the science of natural history. This impression has been further strengthened, by the remarkable fact of no instance having occurred, of late years, of the Copley or any other medals having been bestowed upon any of our naturalists. It seems, however, that this notion is altogether erroneous: for not only does a recent volume of their Transactions contain a zoological paper; but it is expressly stated by the illustrious president, that "physiology, including the natural history of organised beings," holds the second rank in the scale of those sciences, for the promotion of which the royal medals were granted. How great, then, was the astonishment of all those who can rightly appreciate the loftiness of that genius which discovers a law of nature, to see that one of the greatest names in the annals of modern zoology was entirely overlooked in the late distribution of these national medals; while, in order that one should

be bestowed among the cultivators of natural history,
the council deemed it expedient to look abroad for a
fit subject, and to bestow it upon one whose labours
in the minutiæ of his science have been indeed interesting, but whose merits, in comparison to those of
the first discoverer of the system of circular affinities,
are much the same as those which exist between a
numberer of the stars, and the discoverer of the
laws of their motion! Was there no member of the
council present, at this most extraordinary adjudi-
cation, sufficiently acquainted with the philosophy
of natural history, or of those Baconian principles
upon which it should be prosecuted, who protested
against an award so signally unjust to native genius?
We have the pleasure of personally knowing the
amiable and excellent professor at Geneva; and
we are thoroughly convinced, that his surprise at
receiving this medal, knowing and appreciating as he
does, the splendid talents of our countryman, must
have been fully as great as that experienced by the
zoologists of England when first informed of the
event. If the value of a scientific discovery is to be
measured by the universality of its application, by
its effect upon all existing systems, annihilating some,
and breaking up all, by the promulgation of a new
and universal law (the *greatest* of which zoology at
present can boast), then does the discovery in
question leave all others, save one*, at an immeasur-
able distance. We presume not to criticise the de-

---

* We allude to the discovery of the metamorphosis of the *Cirripeda* and of the *Crustacea*, by Thompson, before alluded to, p. 345.

cisions of the council in their other awards, or to hazard an opinion how far the censures passed against it by others are well or ill deserved, for these questions relate to branches of science upon which we are ignorant; but in our own walk we may be allowed to form and to express an opinion, and it is this — that the discovery alluded to (which, we may fairly suppose, was unknown to the council of the Royal Society) is, in natural history, what that of gravitation is in astronomy. If this developement of the first great law of natural arrangement has not yet been seen in its true magnitude, it is because our naturalists, absorbed in the minutiæ of details, shrink from the complicated and severe researches necessary for its verification. We should have rejoiced, had the imperishable fame, which future ages will bestow upon him who achieved so brilliant a generalisation, been anticipated by the Royal Society of Great Britain: yet at the eleventh hour an unintentional act of injustice may still be rectified: and we believe, that the council, upon further enquiry, will admit the validity of our objection. We feel quite satisfied that the illustrious president, no less than his advisers, will not be backward in awarding, upon a future occasion, to the first philosophic zoologist this country has ever produced, that honour which will most assuredly be bestowed upon him by posterity.

(301.) The Linnæan Society, as far as concerns the cultivation of zoology, is the first in Great Britain; whether as regards seniority of date, the scientific rank of its members, or the value of its published Transactions. The unostentatious and regular manner with which its affairs are conducted, gives it an honourable exception from censure, whether private or public.

(302.) There is probably no society in Britain, which, under other regulations, might do so much to restore zoology to her legitimate elevation as the Zoological Society. And yet, as at present constituted, it seems eminently calculated to encourage that superficial and almost useless taste for natural history now so prevalent, and which arises from the custom of regarding it as an amusement rather than as a science. Where there are ample funds, as in the present case, a judicious management may unite, in equal proportions, popular recreation with the encouragement of legitimate science; for the attraction of the former would raise funds for paying the latter, and thus the highest objects might be combined with those that were more ornamental than useful. Our idea of what a society, so constituted, *should* do, is as follows:— Three or four competent persons should be in the regular pay of the society, as travelling naturalists, who should be sent to different parts of the world to collect live animals, and preserve dead ones. Let them be furnished with proper instructions, as to those subjects to which they should more particularly devote their attention, such as the habits and manners of particular species in a state of nature. Their journals should be kept regularly, and transmitted from time to time to the society. To diminish, in some measure, the expense of these missions, the duplicates, of which there would be a large proportion, might be sold by auction for the benefit of the society, or by private contract among the members. This plan would, in all probability, diminish the expense to one half, and the society might have its menagerie recruited by its own officers direct, instead of paying large sums with the accumulated profits of many intermediate dealers; so that finally it might be reasonably expected, that this plan would add but very little to the present expenditure of the society, in purchasing their animals through the instrumentality of several agents. The museum, also, would be thus acquiring a constant accession of new and interesting objects, unprocurable by any other means. In a few years, instead of the present poor collection in Bruton Street, altogether unworthy both of the name and the funds of the society, it might have a museum to which the public would willingly pay for admittance, and feel satisfied with a payment which, at present, is certainly too high. Materials being thus provided, let them be turned to good use. If the society be unwilling to embark in publishing them in a complete and scientific form, let the museum be opened, without vexatious restrictions, to all who are disposed to take such risk or expense upon themselves, no matter whether they be Fellows or not. The very least that can be done, in the way of liberality to scientific men, is, to give them the facility of doing that which the society declines, and which so very few individuals have either the disposition or the talents to accomplish. It savours of that narrow and despicable spirit which is now fortunately so rarely to be met with,—to turn a museum into a scientific preserve, where none but the members are allowed to hunt for information.

If, indeed, there was no deficiency of scientific and working naturalists in the society, who would not suffer these objects to remain in the museum, year after year, unexamined and unrecorded, until time or moths consumed them, then, indeed, the case would be different. But the contrary is notoriously the fact: the museum of this society, under the present regulations, is of little or no use to the science of the country; the members make very little use of it themselves, and prohibit it to others*, who have generally the abilities, and the industry, to turn it to advantage. Short specific characters may do very well for securing the first honours of nomenclature; but this primary examination, after all, is merely skimming the surface of things; and even this, if we are rightly informed, has never yet been done to the museum in question, at least so far as ornithology is concerned, the most inviting branch of vertebral zoology.

(303.) But this negative encouragement of science is not all that would be done by such a society, if it really wished to build itself a solid scientific reputation, apart from that popularity which it will always derive from its gardens and menageries. Might it not be reasonably expected, that from this society a series of illustrative works should emanate, on all new objects coming into its possession? If not, it could at least appropriate an annual sum for subscriptions to such publications of this description as are deemed worthy of support. Works with coloured plates, for instance, which hardly ever

---

* See Fauna Americana Borealis, vol. ii. pref. p. lxii.

pay their bare expenses, should receive its decided patronage, and be regularly subscribed for, as a matter of course. Nay, the society might go a step further, and subscribe for ten or more copies; one of which being deposited in the library, the others might be distributed by lot, or otherwise, among the members, or exchanged with foreign academies, or authors, for similar works, not already in the library.

(304.) The institution of annual prizes or medals is another effectual mode of advancing the true knowledge of zoology, and also of honouring and rewarding its votaries. With such enormous funds at their disposal, why do not the Zoological Society institute two or three annual medals, or premiums, for the best essays upon the innumerable subjects belonging to pure zoological science, now lying open, as a field inviting the reapers, but into which no one will put his sickle! Why this backwardness exists has already been stated. Every writer who courts popular applause must make natural history light and amusing,— or, in other words, treat his subject superficially; and thus the very few among us, who are qualified to extend the boundaries of philosophic zoology, abandon original research, which is neither regarded nor understood, and betake themselves to a less honourable, but more profitable, occupation — the compilation of little volumes, and the editing of animal biographies. What more effectual method, therefore, exists, for raising the tone of the public mind — of withdrawing it from the comparatively trivial and isolated facts of natural history, to its comprehensive sublimities and large generalisations — than the institution of prize essays on the philosophy of the science, given by a society already possessing a popular reputation, and awarded or distributed with all the " pomp and circumstance" appropriate to such occasions? Let these rewards not be merely confined to the members of the society, or even to British naturalists in general — let them bear the stamp and dignity of that enlarged and liberal policy which knows no distinction of persons or of nations, — let them be thrown open to the learned naturalists of all parts of Europe. Let the greatest competition be excited. Let these prizes become objects of ambition, not for their intrinsic value, but for the scientific honour they would confer. Then, and then only, will our zoological institutions advance the true interests of science, instead of limiting its office to the low standard of vulgar minds, and the sole purpose of popular amusement and recreation.

(305.) It is impossible, in a society constituted as this is, to make scientific acquirements a necessary qualification for admission; nor would it, indeed, be at all desirable. It would be quite sufficient if there was but *one* society in Britain where neither money nor interest could procure admission — that should be the Royal Society, or a new one. The Zoological might well continue on their present footing in all things, but those upon which we have just enlarged. By devoting one half, or even one third, of its present revenue to the promotion of true science, there would surely be enough left to purchase amusement for the public; and the society, from being virtually a mere association of amateurs for encouraging the import-ation of living animals, and exhibiting them to the public, would become the powerful and efficient promoter of zoological science, and be honoured and extolled throughout Europe. Opposed, as we always have been, to that illiberal feeling which the influence of one or two of its members have diffused over its councils, we firmly believe that a beneficial change has already begun its work. That when the *substance*, and not the *shadow*, of scientific zoology will be better understood, the Zoological Society will realise all that its real friends and its supposed enemies can possibly desire.

(306.) It is upon this, more than upon any other society, that the benevolent duty devolves of putting aside a small percentage from their funds for decayed naturalists, and their families. Zoological collectors, exploring wild and often unhealthy regions, are exposed, more than any other description of men, to the chances of a premature death. It is fit, therefore, that an association like this should be mindful of men so deserving; and, should they have families, administer to their widows and their orphans some small support out of their abundant wealth. What " golden opinions" might be gained from all men, if the society, for instance, set apart the shillings paid by the visiters to their museum, for the purpose of forming a charitable fund of this description! How cheerfully, for such a purpose, would visitors part with their money; for how nobly would it be appropriated!*

---

* A case of peculiar distress is now before the scientific world. The late Rev. Lansdown Guilding, one of the first

(307.) The Entomological Society, we would fain hope, is too young to have imbibed any thing but zeal and energy in the prosecution of a fascinating science; nor can we foresee but one result, if its proceedings are governed by the general wishes of its members. It has but scanty funds—let them be husbanded well; we hope that associates will be admitted, taken from among the poorer *brethren of the net;* and that the influential members will inculcate a taste for sound philosophic induction, rather than for speculative theories and technical descriptions.

(308.) To the *British Association* for the advancement of science might be suggested a few hints. The plan of forming sections or committees upon the different branches of physical science is admirable, and might be rendered doubly advantageous, were it one of the duties imposed upon the chairman or secretary of each, to draw up a short but comprehensive report on the progress which their own particular science has made during the past year; including, if possible, abstracts of the most important discoveries, and short biographical notices

---

of our zoologists, has been taken from us in the prime of life, leaving a widow and four children totally unprovided for. Had the relict of this distinguished man been *old*, there are public charities for the widows of the clergy, which might have protected her: but because she is *young*, she is excluded from their influence. Can she find no friend among the high and titled members of this society, who will advocate her cause in the proper quarter? Or is there no one, in this country, to befriend " the widow and the fatherless " of an accomplished *savant* ?

of the chief members who have been taken from its ranks. In this respect the speeches or reports of H. R. H. the President of the Royal Society, are admirable models, and deserve to be imitated by every society. The separate reports might then be collected into one, and printed. By thus committing the different branches of science to those most conversant with them, this collection of reports would assume an importance far greater than any other, the work of one or two individuals only, could possibly enjoy. The range of the physical sciences is now so wide, so many discoveries and revolutions are continually going on in each, and the diversified knowledge necessary to appreciate the true value of these changes so vast, as to render it beyond the power of any mind, however powerful, to grasp the whole. This difficulty, however, would be entirely done away with by the plan now suggested. The funds of the society might not allow the institution of prize essays; but it would be highly to the advantage of science if each section proposed, in committee, some one particular subject for research or investigation during the next year; the best essay or paper upon which might be printed at the expense of the society, and some honorary mark of distinction might be conferred upon its author. In natural history, for instance, no subject could be more appropriate than the confirmation or fallacy of any particular theory upon natural affinities; always taking care to select, as the proposed theme, some subject which will call into application those Baconian principles of philosophy upon which all true science must repose.

(309.) We are rejoiced at bringing this part of our subject to a close. Prompted by a disinterested zeal for the advancement of that science to which life has been devoted by us, and anxious that all the societies founded for its advancement should acquit themselves with honour, we have felt it a duty to point out those defects which, as we conceive, prevent their full success. These opinions, it is true, are but those of a very humble individual; yet, as they are founded upon some experience, and upon influences which operate universally, they are not undeserving of attention. Censure, under any circumstances, should be indulgent where the intentions are good. It cannot be supposed, for a moment, that any man, or body of men, having the least love for science, would associate together but for the real purpose of advancing its interests. We may all agree in the object, but differ materially as to the means of accomplishing it. Difference of opinion, therefore, among honest and ingenuous minds, is productive of this good — that new views are elicited, and old ones placed in new and unexpected lights. That freedom of discussion, which, when conducted in a good spirit, is the best safeguard of a government, spreads its beneficial influence over all minor associations, and is a perpetual check upon that tendency to abuse and decay, inseparable from all human institutions.

(310.) Having now suggested the chief of those improvements by which the interests of science can be upheld by our public societies, it is expedient, in this place, to notice a plan which has been talked of among a few of the highest scientific individuals in this country, and whose names alone would ensure success to any measures they deemed it expedient to adopt. This plan consists in the formation of a new society, composed entirely and exclusively of the *élite* of science; and into which no member should be admitted, unless his reputation was already established by his writings, or unless he delivered to the society an original paper, certified as being entirely his sole and whole composition*; his *calibre* would be then known, and his admittance or rejection decided upon by ballot. Associates would also be admitted, chiefly selected from distinguished foreigners: the subscription would be comparatively small, so as not to operate as a pecuniary objection. The number of members would be very limited, so that not more than two, or, at most, three, in each department of the physical sciences, would be admitted. When the society consisted of about thirty or forty, new elections would only take place when vacancies arose from death or otherwise. Such are the main

---

* It may appear singular that such a certificate should be necessary, but the ingenious author of the " Reflections," however ably he has exposed *most* of the frauds of science, seems yet to be unacquainted with one, which has been extensively practised of late years among naturalists. It is for an unscientific individual to get some " friend " to write a paper for a journal or a society, describing his discoveries, and to which his name is appended as *the author*. We know, from personal knowledge, several instances of this fraud. The most remarkable, however, are those that have been practised upon the Linnæan Society; in whose Transactions are two papers on ornithology, bearing the name of one whom we happen to know *can scarcely write his own name*.

features of the contemplated association; and, if its necessity be conceded, it is impossible it can be founded upon better principles. It is urged as a reason for such a society, that no effectual support can be expected from government either in the way of creating a scientific tribunal, whose opinions would deserve the confidence of the nation, or in bestowing titular or honorary distinctions, by which the true philosopher can be recognised from the mere pretender, or the wealthy amateur. It is likewise argued, that the Royal Society, up to the present time, has done nothing to reform its internal government, or to remove the least of those complaints that have been urged against it from so many quarters.* It is further contended, and we think justly, that such a society, by the difficulty there would be of belonging to it, would at once concentrate the united talents of the kingdom, to the exclusion of all common-place merit; and that it would consequently become an object of the highest ambition with men of science, to be enrolled among its members. Whether the meetings were held in a palace or a hovel, would be perfectly immaterial; for all who had scientific ambition would ardently desire to be *of the élite:* they would find that neither titles, nor wealth, nor interest, would avail

---

* It should be stated, however, that full and very satisfactory accounts of the pecuniary transactions of the society have recently been printed and sent to the members. The publication of the president's speeches is also an improvement. But the contemplated plan of recruiting the funds of the society, by increasing the admission fees, is only an aggravation of the evils detailed at p. 434.

them; and they would therefore strive, *by study,* to reach that proficiency which was the sole qualification for admission. That such a feeling would stimulate many of those who are now content with moderate acquirements, is perfectly obvious; and that this study would lead to a great extension of knowledge, is no less evident. We must confess, that if no steps are taken by the present council to raise the scientific character of the Royal Society, or by the government to distinguish in some way our men of science, we cannot but wish to see such an association as the above matured and embodied: and we should desire it upon the ground, that it would effect present and future good; that it would serve as a nucleus for assembling together those few of our first-rate philosophers whose names are now scattered among hundreds of amateurs, in dozens of societies; and that, under the auspices of such men, a tone and a vigour would be given to the science of Britain, which it seems almost hopeless to expect from any other quarter.

APPENDIX.

Letter from the Rev. Thomas Newcome, M.A., Rector of Shenley, Herts, on a Plan for instituting Professorships of Zoology in the English Universities.

Shenley Parsonage, 7th March, 1834.

MY DEAR SIR,

Concurring with you in opinion that the Government of this great country does less for the encouragement of science than that of any other civilised state, I am not disposed to admit that our English universities and ecclesiastical establishment are copartners with our civil governors in the disgrace attaching to them by this statement of a fact. As to " Natural History," as a science, it was not " come to the birth"—was scarce indeed conceived, or in embryo state—at the time when the several colleges were founded, and scholarships and fellowships endowed, by the pious and munificent of days gone by. These men saw and felt the want of something more immediately necessary than science itself; and it is no imputation on their judgment or their charity—on their heads or their hearts—that they provided, in the first place, and by due preference, " for their own." Had they not done so, they would have been "worse than the infidel" of modern times, who endows no institution for the promotion of that science he affects to value as the favourite of Liberals, and the one thing " useful." Considering, however, how delightful a study, and how cheap an amusement is ever within the reach of the " country parson," in any branch of natural history, such as a *Herbert* might recommend to him by precept, and a White of Selborne or a Kirby by example,—such, too, that no Squire can grudge the Rector his field sports in this kind, nor Bishop will deny or discourage their prosecution,—that *natural* religion is the basis of *revealed*, and is itself built on some observation of nature*, I would propose the endowment of a Pro-

---

* Romans i. 20.   Psalm xix. 1.   Acts xiv. 17. and xvii. 17.

fessor of Zoology in Oxford and Cambridge, whose public lectures would at least serve "just to put the taste for this study into the mental *mouths* of the young disciples, to be followed up when they shall have quitted college." The stigma on the Government, and on the *lay members* of the universities, for thus long neglecting to promote this science, may be cheaply redeemed by the former remitting its abominable stamps upon Degrees taken in Arts, viz. three guineas, I believe, upon the A.B., and six on the M.A. degree. These sums are paid by hundreds of clerical parents, who at great cost and homefelt sacrifice educate one son for the church. Now these are taxes, doubly and trebly faulty, as being *partial*, if not oppressive; and, unlike the newspaper tax, are obstacles in the road to obtaining that "sound learning and religious education," and that "useful knowledge," also, which colleges profess and are intended specially to promote, "for fit supply to serve God in Church and State."* I have not the means at hand of estimating correctly the produce of these taxes; but, from an old Cambridge Kalendar of the year 1827, I learn that about 200 persons then took the degree of M.A. This gives 1200 guineas for *one* degree, in *one* only of our universities. So that, for all degrees in Oxford, Cambridge, and Dublin, I will assume 5000 is annually paid into the national treasury, which repays back less than one fifth, or 1000*l.*, to ten Professors in Cambridge. This is the sum total of national help to that university. Here, then, is a fund for endowing ten professorships of 500*l.* per annum, on any of the more neglected sciences,—"a refuge for the destitute,"— such as many an able and languishing son of science would gladly accept, and for which he would fearlessly and cheerfully devote his services. Let the Government, then, remit the tax, but not, as now, pay and appoint the professors. Let all the graduates pay the same sums as before, but into the university chest: and, so paying, let all (that will) exercise a right of voting in person or *by proxy* for any candidate, whether he be a member or not, at the time, of the university. Let these Professors be actual masters and lecturers in their science — not sinecure dignitaries of the scientific world, but labourers

---

* Quoted from Canonical prefatory Prayer before Assize and Visitation Sermons, &c.

worthy of their hire, and above " carking care" for bread. I
think they should not be bound to take holy orders, nor yet
prevented from so doing. At the same time, I do not think
that these new professors for the *university at large* should have
a share in the internal government of any particular college.
This would preclude any distraction from their proper province
— might open a safe door for the scientific dissenter — yet reserve to the clerical members of the colleges that prescriptive
right which they now possess in their discipline and emoluments; it being the evident, if not expressed, intention of the
founders to make the MAIN object of our university education
the preparing men to serve God and man in the government
of the Church and State. Science that is not honourably subordinate to these great and nobler ends, and directed Heaven-
ward, is mere temporary " *utilitarianism* ;" — a degraded handmaid to personal pride, to pecuniary profit, and to corporeal
ease, — a thing, in short, of *this* world only, and to perish with
its other vanities. I am,

My dear Sir,
Your true, though not scientific Friend,
THOMAS NEWCOME.

P. S. — Since writing the above, I have been able to procure the following document, which at once substantiates
my argument, and shows to what extent the universities
administer to the expenses of that government which should
support *them*.

A RETURN made to Parliament of the number of members
admitted to the two Universities, and the degrees granted
by the same, in each of the three years 1831, 1832, and
1833, with the amount of duty on each degree, and the
aggregate amount of each year : —

The number of Noblemen and Fellow Commoners admitted
into the University of Cambridge from the 10th of October,
1830, to 1831, 31; Pensioners, 377; Sizars, 45: total, 453.
1831 to 1832 — Noblemen and Fellow Commoners, 33; Pensioners, 335; Sizars, 41: total, 409. 1832 to 1833 — Noblemen and Fellow Commoners, 48; Pensioners, 345; Sizars,

38: total, 431. Each person upon his matriculation pays the sum of 1*l.* to Government. The degrees conferred from October, 1830, to 1831 — D.D., 8; D.C.L., 1; D.M., 5; B.D., 13; B.C.L., 9; B.M., 8; Licentiate to practise Medicine, 4; M.A., 205; B.A., 323: total, 2535. From 1831 to 1832 — D.D., 3; D.C.L., 1; D.M., 3; B.D., 10; B.C.L., 12; B.M., 10; Licen. to prac. Med., 3; M.A., 185; B.A., 316: total, 2,334. From 1832 to 1833 — D.D., 2; D.C.L., 6; D.M., 3; B.D., 15; B.C.L., 13; B.M., 9; Licen. to prac. Med., 3; M.A., 213; B.A., 302; Mus. Bac., 1: total, 2558. Each person on his admission to B.A. pays to Goverment 3*l.*; to any other degree, 6*l.* There are in each year a few noblemen, each of whom, upon his admission to any degree higher than B.A., pays 10*l.*

The number of degrees granted in the Oxford University in 1831: — D.D., 6 at 6*l.*; D.C.L., 2 at 6*l.*; D.M., 1 at 6*l.*; B.D., 8 at 6*l.*; B.C.L., 7 at 6*l.*; B.M., 1 at 6*l.*; M.A., 177 at 6*l.*; B.A., 268 at 3*l.*; B. Mus., 1 at 3*l.* — Total number of degrees, 471; total amount, 2019*l.* Certificates of degrees — 10 at 10*l.*, and 1 at 3*l.* — 103*l.* Matriculations, 380 at 1*l.* — 380*l.* Grand total for the year 1831, 2502*l.*

In 1832: — D.D., 2 at 6*l.*; B.D., 8 at 6*l.*; B.C.L., 4 at 6*l.*; B.M., 1 at 6*l.*; M.A., 175 at 6*l.*; B.A., 270 at 3*l.* Incorporations — M.A., 1 at 6*l.*; B.A., 2 at 3*l.* — Total number of degrees, 460; total amount, 1962*l.* Certificates of degrees, 18 at 10*l.*, 6 at 3*l.* — 198*l.* Matriculations, 393 at 1*l.* — 393*l.* Grand total for the year 1832, 2553*l.*

In 1833: — D.D., 4 at 6*l.*; D.C.L., 3 at 6*l.*; D.M., 3 at 6*l.*; D. Mus., 1 at 6*l.*; B.D., 10 at 6*l.*; B.C.L., 1 at 6*l.*; B.M., 5 at 6*l.*; M.A., 185 at 6*l.*; B.A., 293 at 3*l.* Incorporation — B.A., 1 at 3*l.* — Total number of degrees, 507; total amount, 2160*l.* Certificates of degrees, 10 at 10*l.* — 100*l.* Matriculations, 363 at 1*l.* — 363*l.* Grand total for the year 1833, 2623*l.*

INDEX.

ACADEMY of Science at Petersburg, 351.
Adanson, remarks on his works upon botany and zoology, 79.
Ælian, 10.
Affinities, discovery of the circular nature of, 91.
Affinity and analogy, doctrines of, 116.
Albin, 32.
Aldrovandus, Ulysses, the naturalist, remarks on his works, 14.
Analogy and affinity, theoretical distinction of, 214.
Analogy, importance of, when applied to the confirmation of theory, 282. Analogy between the natural and the moral world, 283. Importance of, to natural history, 284. Difference between an argumentative and an illustrative analogy, 286. Interest arising from, 289. General effects and advantages produced by, in the elucidation of truth, 290. Three sorts of analogies, 291. Material and spiritual, 293. Applicable to physical science, 295.
Animals, form and construction of, 167. External distinction preferable to internal, 169. Internal construction of, 171. Aids afforded by anatomy, 172. The properties of, 173. Habits and economy of, 174. Diversity of the habits and operations of, 175. Properties of, in regard to their influence or uses in the economy of nature, 179. Instances of analogy and affinity of, 183. Contrariety of structure in, illustrated, 229. Gradation of form in, 231. Results of numerical equality in tribes, 233. Characters of natural groups, 236. Universal characters objectional, 239. Variations of character, 241. Generic characters of, 243. Uniformity in natural groups, 245. Essential characters of, 248. Simplicity of definition, 249. General form of, 251. Appendages of the head, 253. Characters from caudal appendages, 255. Characters of, founded on the structure of the mouth, 259. Suctorial animals, 263. Value of distinctions derived from the organs of locomotion, 264. Progression of molluscous ones, 275.
Aristotle, 5.

INDEX.

Arrangement of insects, 201. Difficulties of, 203. Groups not always perfect, 213. Order of succession, 215. Verification of groups, 217. Importance of uniform results, 223.
Art and nature, reflections on, 97.
Artedi, one of the earliest and most eminent disciples of the Linnæan school, 45. Remarks on his writings, 46.
Ashmolean Museum presented to the university of Oxford in 1682, by Dr. Elias Ashmole, account of, 320.
Asiatic Society, the, 330.
Audebert, 89.
Author, anecdote of the, 129.
Azara, Don Felix de, the Spanish naturalist, 81. Remarks on his writings, 82.

Babbage, Mr., 306.
Bacon, 7.
Banks, Sir Joseph, 39. Anecdote of, 128.
Barbut, 70.
Barraband the best ornithological painter France ever produced, 81.
Bat, various opinions of the individual construction of the, 153.
Bauer, F., 350.
Beechey, Captain, 383.
Belon of Mans, 10. Observations on his works, 11. View of his system, 12.
Belzoni, the celebrated Egyptian traveller, 393.
Berzelius, 352.
Bicheno, Mr., 364.
Birds, diversity of habits among, 177. Their feet the means of discriminating the primary divisions of the feathered race, 266. Feet of climbing birds, 269. Claws of birds, 271.
Bloch, remarks on his famous work upon ichthyology, 61.
Boccone, the famous Sicilian botanist, 24.
Boerhaave, 22.
Bonnet, 21.
Bontius, 19.
Borlase, 50.
Born, remarks on his work upon systematic conchology, 59.
Bradley, 32. 70.
Brisson, M., remarks on his writings, 77.
British Association for the Advancement of Science, 327. Objects of, 328. Suggestions for the improvement of, 445.
British Museum, 332. Recent national encouragement given to, 337.
Brown, Mr., 32.
Bruguire, M., 161.
Buffon, 44. Remarks on his works, 45. Character and progress of his school, 74.

Cambridge Philosophical Society, 318.
Catesby, 33.
Catherine II. of Russia, 59.
Chain of being, 205. Paucity of its known laws, 207.
Colonna, Fabius, remarks on his treatises upon natural history, 15.
Contrariety of opinion on the structure of animals, 153.
Continuity of structure of animals illustrated, 153.

INDEX.

Cook, Captain, 55.
Copyright act, 398.
Cramer, remarks on his writings, 57.
Croonian lecture, origin of, 310.
Cuvier, M., 81. Remarks on his work entitled *Règne Animal*, 86.

D'Argenville, remarks on his work entitled *Conchology*, 41.
Daubenton, 44.
Davy, Sir H., 340.
Denon, the celebrated Egyptian traveller, 393.
Desmarest, 89.
Different analogies in the creation, 115.
Dillwyn, Mr., remarks on his conchological writings, 73.
Donovan, remarks on his writings, 70.
Drury's *Exotic Insects*, 52.
Duhamel, remarks on work upon ichthyology, 80.
Duncan, Mr., 323.

East India Company, the Hon., 329.
Edwards, George, 41. Remarks on his writings, 42.
Elgin Marbles, the, 97.
Ellis, John, immortalised by his discovery of the nature of coralline animals, 38. Remarks on his writings, 39.
Engramelle, father, an Augustine monk, remarks on his work on the Lepidoptera of Europe, 68.
Entomologic student, dissections for the, 201.
Entomological Society, foundation of, 317. Suggestions for the improvement of, 445.

Entomology, 46. Rise and progress of, 47.
Ernst, 68.
Esper, his work upon the Lepidoptera of Europe, 68.
Eudamus, a species of butterfly, 210.

Fabricius, Otho, remarks on his writings, 53. His work on the Zoology of Greenland, 67.
Fairchild's lectures, origin of, 308.
Feræ, 266.
Fermin, 32.
Fischer, Professor, the celebrated zoologist of Russia, remarks on his writings, 91.
Flinders, Captain, 350.
Forester, Mr., 55.
Forskal, M., 49.
French school of zoology, 82.
Fries, E., his discovery of the circular nature of affinities in the vegetable world, 92.
Fuessly, J. G., remarks on his writings, 82.

Garden, Dr., 60.
Geer, Baron de, remarks on his writings, 77.
Genera, 211. Extensive genera favourable for study, 219.
Geological Society, 313.
Gionea, the genus, origin of, 161.
Gmelin, Dr., remarks on his writings, 73.
Goderich, Lord, 337.
Goedartius, his experiments in entomology, 20.
Gradation of form, 231.
Grew, Dr., remarks on his works, 31.

Gronovius, remarks on his writings, 43.

Hampden, extracts from his work on the truths of natural history as connected with those of religion, 291.
Hanstein, Professor, 352.
Hardwicke, General, 312.
Harris, Moses, 52.
Hasselquist, 49.
Hawk, the, 146.
Haworth, Mr., extract from his work entitled Lepidoptera Britannica, 126.
Hedgehog, experiments to ascertain the natural food of the, 160.
Herbert, Mr. Thomas, 322.
Herbst, his work upon entomology, 69.
Hermann, Professor, remarks on his writings, 60.
Herschel, Sir John, 340.
Hessian fly, panic caused by the supposed appearance of it in this country, 140.
Honorary titles, uses of, 417.
Horæ Entomologicæ, 91.
Horsfield, Dr., 331.
Hübner, his works on entomology, 69.
Humboldt, Baron, 349.

Ichthyology, rise and progress of, 61.
Illiger, 89.
Illustrative works, sale of, 401. Importance of, 403. Suggestions for facilitating the circulation of, 404.
Insects, feet and wings of, 273. Remarks on the metamorphosis of, 276.

Jablonsky, his work on entomology, 69.
Jacquin, the celebrated botanist, 350.
Jay, the, 145.
Jones, Sir William, 330.

Kirby, Mr., 88.
Klein, 32.
Knorr, 32.

La Cépède, 61.
Lamarck, M., 83.
Latham, Dr., remarks on his *General Synopsis of Birds*, 63.
Latreille, M., 55. 83.
Laspeyres, remarks on his entomological works, 72.
Le Vaillant, 80. Remarks on his works upon ornithology, 81.
Leach, Dr., 335.
Linnæan school, rise and progress of, 44.
Linnæan Society, 312. Suggestions for the improvement of, 439.
Linnæus, Sir Charles, 5. Remarks on his character and writings, 34.
Lion, analogy of the, with the vulture, 185.
Lister, Dr. Martin, 10. Remarks on his writings on conchology, 23.
London University, 324.
Lyonnet, 21.

MacLeay, Mr., his discovery of the circular nature of affinities, 92. His illustration of the continuity of the structure of animals, 228. His remarks on the metamorphosis of insects, 277.

INDEX. 459

MacLeay, Alexander, 312.
Man, design in the creation of, 110. Unnecessary to the operations, and disconnected with the designs, of nature in the material world, 111. The apparent anomaly of the design in the creation of man explained by an enquiry into the truths of religion, 112. Inferences from design, 113.
Manchester Natural History Society, 325.
Marcgrave, remarks on his works upon zoology, 19.
Marco Polo, 147.
Marsham, 70.
Martini, his great work on general conchology, 52.
Martyn, 70.
Maurice of Nassau, 17.
Merram, remarks on his writings, 60.
Merrett, Dr., his *Pinax* the first work that was devoted exclusively to the animals and plants of Great Britain, 19.
Mouffet, his *Theatrum Insectorum* the first zoological work ever printed in Britain, 15.
Müller, 55.

Natural history, state of, in the early ages of the world, 5. Declension of, under the Romans, 9. Introductory remarks on the study of, 93. Its general nature and advantages, 95. Writings of the ancients on, 97. Striking advantages in reference to the human mind resulting from the study of, 100. Distinctions and objects of the study of, 102. In its early stages, a science of observation; in its latter, one of demonstration, 105. Its connection with religion, 107. Viewed as a recreation, 117. The study of, congenial to a country life, 118. Subservient to the economic purposes of life, 122. A relaxation from business, 123. Conducive to health, 125. Beneficial to invalids, 127. Reflections on the study of, 130. Considered in reference to commerce, and the economical purposes of life, 133. In reference to its intimate connection with agriculture, 139. Acquaintance with, useful to planters and emigrants, 142. A knowledge of, an essential qualification for travellers, 147. As a philosophical study, elaborate and difficult, 150. Dismissal of prejudice absolutely essential to the study of, 152. A science of facts and inferences, 153. Exempt from general laws, 156. Necessity of correcting the prejudices of sense in the study of, 158. General directions for the study of, 201. Its special claims for support on the national institutions, 356. Expense of naturalists' materials, 359. Sublime and pleasurable sensations resulting from the study of, 375. The study of, recommended at our universities, 377. Objections answered, 379. Its

advantages in after-life, 380. The study of, calculated to advance the interests of religion, 382. Neglected by the government, 383. How to be encouraged by the government, 385.
Naturalists, sent on voyages of discovery, 393.
Newcome, Rev. J., his letter upon the tax on universities, 386.
Newton, Sir Isaac, 7.

Occupation of scientific men, 363.
Olivier, the celebrated French entomologist, remarks on his work upon coleopterous insects, 56.
Olivi, his work on the marine productions of the Gulf of Mexico, 67.
Ornithology, 62. Rise and progress of, 63.
Ortus Sanitatis, the first printed work that treated on the nature of animals, 10.
Osbeck, 49.

Pachydermata, 252.
Pallas, Professor, 57. Remarks on his writings, 58.
Panzer, remarks on his entomological works, 71.
Payhull, remarks on his entomological works, 71.
Penn, Thomas, 16.
Pennant, remarks on his writings, 50.
Petagni, remarks on his entomological works, 71.
Petiver, 31.
Phanæus, 216.
Pinto, Ferdinand Mendez, 147.
Piso, 19.

Pliny the elder, remarks on his writings 8.
Polecat, the, 145.
Poli, 87. Remarks on his work upon the comparative anatomy of the mollusca, 89.

Raffles, Sir Stamford, 17.
Rawlinson, Dr., 322.
Ray, the naturalist, 10. Remarks on his writings, 29.
Reaumur's Memoirs towards the History of Insects, 43.
Redi, M., his experiments in entomology, 20.
Regenfuss, 41.
Religion, connection between natural and revealed, 372.
Renard, 32.
Robin redbreasts, the natural history of the, 144.
Roemer, 24. 56.
Rœsel, 41.
Rondeletius, remarks on his work upon ichthyology, 12.
Rosse, Professor, 71.
Royal Society, 311. Suggestions for the improvement of, 430.
Royal Museum at Munich, 351.
Rumphius, remarks on his works upon conchology and botany, 40.

Salviani, remarks on his work upon ichthyology, 13.
Savigny, 87.
Schellenberg, remarks on his work upon the two-winged genera, 70.
Schneider, Professor, his writings, 62.
Schoeffer of Ratisbon, remarks on his writings, 48.
Schrank, his works, 55.

Schreber, remarks on his work upon quadrupeds, 57.
Schroeter, remarks on his work upon systematic conchology, 59.
Science, present state of, in Britain, 339. Claims of, for patronage from the government, 342. Diffusion and extension of, in England, as compared with other countries, 343. Neglect of comprehensive enquirers, 345. Continental patronage, 347. By whom it should be protected, 355. Concluding reflections in reference to all the physical sciences, 365. Means possessed by the government and universities for encouraging it, 367. Neglected at our universities, 371. Consequences of the decline of, 427.
Scientific societies, influence of, 297. Present state of, in Britain, 299. Publishing committees, 305. Rewards and medals given by, 306. Difficulty of filling a scientific order, 419. Distinctions should be made, 421. True cause ministerial neglect, 423. Suggestions for the improvement and reform of, 428. Outlines of a new one, 449. Scientific men, appropriate offices for, 389. Propriety of conferring honorary titles on, 407. Scientific noblemen in France, 409. Honours withheld from Englishmen, 412. Discussion of the subject in parliament, 414. Objections answered, 415.

Scopoli, remarks on his work upon the entomology of Carniola, 47.
Scotch universities, 324.
Seba, Albertus, 33.
Sepp, remarks on his works on the insects of the Low Countries, 47.
Shaw, Dr., remarks on his writings, 65.
Sloane, Sir Hans, remarks on his works, 31.
Smeathman, Mr. Henry, his interesting account of the insects generally termed white ants, 52.
Smith, Sir James, remarks on his works, 66.
Solander, Dr., 39.
Sonnerat, his works, 80.
Sonnini, his works, 80.
South, Sir James, 429.
Spallanzani, 48.
Sparmann, 49.
Sparrows, natural history of, 145.
Squirrel, the, 145.
St. Hilaire, Geoffroy, his system of ornithology, 87.
Stewart, Dugald, his opinion of the importance of analogy, 284.
Swammerdam, his laborious researches and anatomical discoveries, 21.
Systematists, prejudices of, 157.
Systems, natural and artificial, 188. Advantages of artificial ones, 191. Origin of mixed systems, 197. Natural systems alone conducive to the advancement of natural history as a physical science, 200. Inconsistency of artificial ones, 237.

Temminck, M., 81.

Theatrum Insectorum, the first zoological work ever printed in Britain, 15. Remark on, 16.
Theories in general, 201. Modes and considerations by which they are to be verified, 206. Importance of analogy when applied to the confirmation of, 282.
Thunberg, 55.
Topsel, Edward, his works, 17.
Tradescant, John, 321.
Trembley, 38. His discovery of the reproductive powers of the freshwater polype, 42.
Trichius, the genus, 216.
Trochilidæ, 224.

Uddman, 70.
Universities, tax upon, 386.

Viellot, 89.
Villiers, 55.

White of Selborne, 51.
Wilks, Benjamin, his entomological figures, 52.
Willughby, the naturalist, 10. Remarks on his character and writings, 26.
Wolff, remarks on his work upon European Hemiptera, 70.
Wollaston, Dr., 431.
Woodpecker, the, 145.
Wotton, Edward, 16.
Wren, Sir Christopher, 431.

Zoological Society, constitution of, 314. Suggestions for the improvement of, 439.
Zoology, general observations on the rise and progress of, 2. Division of the subject, 5. Progress of, in England and France, 88.

THE END.